Food Consumption

Food consumption patterns and practices are rapidly changing in Asia and the Pacific, and nowhere are these changes more striking than in urban areas. This book brings together researchers from environmental studies, anthropology, sociology, tourism, architecture and development studies to provide a comprehensive examination of food consumption trends in the cities of Asia and the Pacific, including household food consumption, eating out and food waste.

The chapters cover different scales of analysis, from household research to national data, and combine different methodologies and approaches, from quantifiable data that show how much people consume to qualitative findings that reveal how and why consumption takes place in urban settings. Detailed case studies are included from China, India, Japan, Malaysia, Philippines, South Korea and Vietnam, as well as Hawai'i and Australia. The book makes a timely contribution to current debates on the challenges and opportunities for socially just and environmentally sound food consumption in urbanizing Asia and the Pacific.

Marlyne Sahakian is a Senior Researcher in the Faculty of Geosciences and the Environment at the University of Lausanne, Switzerland.

Czarina Saloma is Associate Professor of Sociology at Ateneo de Manila University, Philippines, and recent Alexander von Humboldt Foundation Fellow at Bielefeld University, Germany.

Suren Erkman is Associate Professor in the Faculty of Geosciences and the Environment at the University of Lausanne, Switzerland, where he manages the Industrial Ecology Group.

Routledge Studies in Food, Society and the Environment

Civic Engagement in Food System Governance
A comparative perspective of American and British local food movements
Alan R. Hunt

Biological Economies
Experimentation and the politics of agri-food frontiers
Edited by Richard Le Heron, Hugh Campbell, Nick Lewis and Michael Carolan

Food Systems Governance
Challenges for justice, equality and human rights
Edited by Amanda L. Kennedy and Jonathan Liljeblad

Food Literacy
Key concepts for health and education
Edited by Helen Vidgen

Sustainable Urban Agriculture and Food Planning
Edited by Rob Rogemma

Transforming Gender and Food Security in the Global South
Edited by Jemimah Njuki, John R. Parkins and Amy Kaler

Urban Food Planning
Seeds of Transition in the Global North
Rositsa T. Ilieva

Food Consumption in the City
Practices and patterns in urban Asia and the Pacific
Edited by Marlyne Sahakian, Czarina Saloma and Suren Erkman

For further details please visit the series page on the Routledge website: http://www.routledge.com/books/series/RSFSE/

Food Consumption in the City

Practices and patterns in urban Asia and the Pacific

Edited by Marlyne Sahakian, Czarina Saloma and Suren Erkman

First published 2016
by Routledge

2 Park Square, Milton Park, Abingdon, Oxfordshire OX14 4RN
711 Third Avenue, New York, NY 10017

Routledge is an imprint of the Taylor & Francis Group, an informa business

First issued in paperback 2018

Copyright © 2016 Marlyne Sahakian, Czarina Saloma and Suren Erkman, selection and editorial material; individual chapters, the contributors

The right of the editors to be identified as the authors of the editorial material, and of the authors for their individual chapters, has been asserted in accordance with sections 77 and 78 of the Copyright, Designs and Patents Act 1988.

With the exception of Chapter 3, no part of this book may be reprinted or reproduced or utilised in any form or by any electronic, mechanical, or other means, now known or hereafter invented, including photocopying and recording, or in any information storage or retrieval system, without permission in writing from the publishers.

Chapter 3 of this book is available for free in PDF format as Open Access at www.tandfebooks.com. It has been made available under a Creative Commons Attribution-Non Commercial-No Derivatives 3.0 license.

Notice:
Product or corporate names may be trademarks or registered trademarks, and are used only for identification and explanation without intent to infringe.

British Library Cataloguing-in-Publication Data
A catalogue record for this book is available from the British Library

Library of Congress Cataloging in Publication Data
A catalog record for this book has been requested

ISBN: 978-1-138-12061-7 (hbk)
ISBN: 978-0-367-02974-6 (pbk)

Typeset in Bembo
by Saxon Graphics Ltd, Derby

Contents

List of illustrations — vii
List of contributors — ix

Introduction — 1
MARLYNE SAHAKIAN, CZARINA SALOMA AND SUREN ERKMAN

PART I
Trends across the region — 21

1 Same, same – but different: on increasing meat consumption in the Global South — 23
HELLMUTH LANGE

2 Grappling with impacts: environmental considerations around agriculture and food waste treatment in Asian cities — 46
MEGHA SHENOY

PART II
Food practices and cultures — 69

3 Emerging consumerism and eating out in Ho Chi Minh City, Vietnam: the social embeddedness of food sharing — 71
JUDITH EHLERT

4 Eating in vertical neighborhoods: food consumption practices in Metro Manila condominiums — 90
CZARINA SALOMA AND ERIK AKPEDONU

5 From beef to bananas: consumer preferences and local food
  flows in Honolulu, Hawai'i   107
  ELIZABETH LOUIS AND KYLE DATTA

6 The food revolution in Melbourne, 1980–2015   124
  WARWICK FROST AND JENNIFER LAING

7 The practices of Indian vegetarianism in a world of limited
  resources: the case of Bengaluru   141
  RUNE-CHRISTOFFER DRAGSDAHL

PART III
Food waste dynamics   **159**

8 Uneaten food: emerging social practices around food waste
  in Greater Tokyo   161
  ATSUSHI WATABE, CHEN LIU AND MAGNUS BENGTSSON

9 From thrift to sustainability: the changing table manners of
  Shanghai's food leftovers   180
  DUNFU ZHANG

10 Food waste in the food service sector: a case study from
   Kuala Lumpur, Malaysia   199
   EFFIE PAPARGYROPOULOU

11 Convenient food, inconvenient waste: systems of provision
   meet social practices in Seoul   216
   KEITH LEE

12 Towards sustainable consumption of rice in a private
   school in Metro Manila   238
   ABIGAIL MARIE T. FAVIS AND RAFAEL DEO F. ESTANISLAO

   Index   254

Illustrations

Figures

1.1	Meat consumption, 2011/12–2023: selected countries; per capita kg/year	25
1.2	Meat consumption, 2013–2023: selected countries/types of meat per capita kg/year	26
1.3	Fish consumption, 2013–2023: selected countries; per capita kg/year	27
1.4	Consuming meat and fish in *rural* India; per capita gram/year	32
1.5	Consuming meat and fish in *urban* India; per capita gram/year	33
1.6	Consuming meat and aquatic products: rural and urban China, 2009; per capita kg/year	35
1.7	Frequencies of cooked meals purchased; per capita/month (India)	37
1.8	Consumption of meat and fish in China, 2013–2023; kg per capita/year	39
2.1	LCA inputs and outputs for agriculture and food waste treatment	48
2.2	Low to high impacts of five food-waste processing technologies	57
6.1	A hole in the wall café with grungy/hipster vibe in Hosier Lane	129
6.2	Hosier Lane combines street art, restaurants and cafés and has become an iconic tourist attraction	130
8.1	Food supply in Japan by main category, 1960–2013	163
8.2	Household consumption expenditure for food in Japan, 1980–2011	164
8.3	Generation and treatment of food waste (including edible food waste) in Japan, April 2011 to March 2012	167
8.4	Calorie supply and actual daily individual intake, 1970–2012	168
10.1	Daily food waste generation	204
10.2	Avoidable and unavoidable food waste as a percentage of total food waste	207

10.3	Average food waste generation per customer served for different mealtimes	208
10.4	Material flows	209
10.5	Material flows in terms of food commodities	209
10.6	Economic flows	210
10.7	Eco-efficiency of food commodities	210
11.1	Per capita per day food waste in Seoul, 2004–2013	220
11.2	Hypermarket growth, 1993–2011	221
11.3	Percentage of respondents (aged 10+) who reported spending time shopping for household-related purposes at different times on weekends	223
11.4	Percentage of respondents (aged 10+) who reported spending time shopping for household-related purposes on weekdays	224
11.5	Percentage of married (M) and single (S) respondents (aged 20+) who spent time shopping for household-related purposes on weekdays	225
11.6	Percentage of married (M) and single (S) respondents (aged 20+) who spent time shopping for household-related purposes at different times on weekends	226
11.7	'Convenience vegetables' and prepackaged salads on sale at Homeplus	228
12.1	The food waste collection bin and the plastic trash bins	243
12.2	The university canteen	245

Tables

1.1	Meat consumption: developed/developing countries, 2012/14–2024	25
1.2	Meat consumption in India, 2013, types of meat	29
1.3	Meat consumption in China, 2013, types of meat	30
6.1	Apparent per capita food consumption for Australia, year ended 30 June	126
6.2	Population of Greater Melbourne by ancestry	132
10.4	Daily number of customers and food waste per customer rates	204

Contributors

Erik Akpedonu, Research Associate at the Institute of Philippine Culture of Ateneo de Manila University, studied Architecture at the Lippe University of Applied Sciences in Germany. He is co-author of *Endangered Splendor: The Architectural Heritage of Manila* (Ateneo de Manila University Press, forthcoming) and *Casa Boholana: Vintage Houses of Bohol*.

Magnus Bengtsson is Principal Policy Researcher at the Institute for Global Environmental Strategies (IGES) based in Japan. He is currently in charge of institute-wide flagship initiatives and was previously Head of Sustainable Consumption and Production Research. He has a PhD in Environmental Systems Analysis from Chalmers University of Technology in Göteborg, Sweden, and was a post-doctoral fellow at the University of Tokyo before joining IGES.

Kyle Datta is general partner of Ulupono Initiative, a Hawai'i-focused social investment organization dedicated to improving the quality of life for island residents by creating renewable energy, increasing locally produced food and reducing waste. Formerly, Kyle was CEO of U.S. Biodiesel Group and served as Managing Director of Research and Consulting at the Rocky Mountain Institute, where he co-authored *Winning the Oil End Game* and *Small is Profitable*.

Rune-Christoffer Dragsdahl is a Phd Fellow at the Department of Cross-Cultural and Regional Studies at University of Copenhagen (2014–2017). His current research is on agricultural connections between India, Mozambique and Ethiopia. He was an Assistant Lecturer in Anthropology at University of Copenhagen (2012–2014) and External Lecturer in Food Systems at Danish Institute for Study Abroad (2013–2014).

Judith Ehlert is a sociologist and holds a postdoctoral position in Development Sociology at the University of Vienna's Department of Development Studies. She is the leader of the research project 'A Body-Political Approach to the Study of Food: Vietnam and Global Transformations', which is funded by the Austrian Science Fund (FWF). Besides food, identity and body politics, her research interests include human–nature interfaces and the sociology of knowledge.

Suren Erkman is Associate Professor in the Faculty of Geosciences and the Environment at the University of Lausanne, where he manages the Industrial Ecology Group. He teaches and publishes around the themes of sustainable resource management, eco-industrial development, industrial symbioses and eco-innovation, with recent work on food consumption and energy transitions.

Rafael Deo F. Estanislao graduated with a degree in Health Sciences, focused on Development Management, from the Ateneo de Manila University. He has experience in health systems research and is an instructor at the Health Sciences Program. His research interests include climate change adaptation in health systems.

Abigail Marie T. Favis obtained an MSc in Environmental Sanitation at the University of Ghent, Belgium. Currently she is a faculty member of the Department of Environmental Science, Ateneo de Manila University, and is the programme manager for campus sustainability at the Ateneo Institute of Sustainability.

Warwick Frost is Associate Professor in Tourism, Hospitality and Events at La Trobe University, Melbourne, Australia. He is the editor of *Wine and Identity: Branding, Heritage, Terroir* (Routledge 2014) and co-author of *Gastronomy, Tourism and the Media* (Channel View 2016).

Jennifer Laing is Associate Professor in the College of Arts, Social Sciences and Commerce at La Trobe University, Melbourne, Australia. Together with Dr Warwick Frost, Jennifer is a foundation co-editor of the Routledge *Advances in Events Research* series. Her research interests include travel narratives and tourism and the media.

Hellmuth Lange trained as a sociologist and political scientist. He served as Director of the Centre for Sustainability Studies (artec) at the University of Bremen, and as a visiting professor in India and China. His general research areas are the sociology of science and risk, and environmental sociology. In recent years, he has focused on understanding sustainability regarding both individual preferences and behaviour, and change management as a challenge to conducive governance and policies.

Keith Lee is a doctoral candidate in City and Regional Planning at the University of California, Berkeley (at the time of writing), whose dissertation investigates the linkages between food retail development and food waste in

Seoul, South Korea. His broader interests encompass the effect of urban development on urban metabolism in East and Southeast Asia, focusing on the flows of food consumption and waste. As an interdisciplinary scholar, Keith's work draws upon approaches and concepts from planning, industrial ecology, sociology and geography.

Chen Liu is a researcher at the Institute for Global Environmental Strategies (IGES) in Japan. Her current research is mainly on 3R policies and municipal solid waste management in Asian and Pacific countries. She earned a PhD in Biosphere Informatics from Kyoto University in Japan, worked as a researcher at the National Institute for Environmental Studies (NIES) and is Associate Professor in Nagoya University Graduate School of Environmental Studies in Japan before joining IGES.

Elizabeth Louis is Senior Research and Evaluation Specialist at Landesa, an organization working on land rights for the rural poor. She earned a PhD in Geography from the University of Hawai'i at Manoa. Her research interests include the political ecologies of land, alternative agriculture and sanitation in South Asia and the Pacific.

Effie Papargyropoulou is Visiting Lecturer in Environmental Management at Universiti Teknologi Malaysia (UTM). She is interested in the conflict between development and sustainability, as manifested in different contexts such as food consumption and carbon emissions in Asian cities. She is currently investigating food waste generation and prevention in the hospitality sector in Malaysia.

Marlyne Sahakian is a Senior Researcher in the Faculty of Geosciences and the Environment at the University of Lausanne. Her research interest is in understanding natural resource consumption patterns and social practices in relation to environmental promotion and social equity, and identifying opportunities for transitions towards more 'sustainable' societies.

Czarina Saloma is Associate Professor of Sociology at Ateneo de Manila University and recent Alexander von Humboldt Foundation Fellow at Bielefeld University. Her publications examine the built environment, particularly new spaces and expanding social practices and patterns, and knowledge mobilization, especially the scaling-up processes of social development initiatives.

Megha Shenoy is an independent industrial ecologist. Her research work includes investigating strategies for sustainable food consumption, industrial ecology in developing countries and industrial symbiosis. After completing a post-doctoral fellowship at the Center for Industrial Ecology, Yale University, she led the research program of the Resource Optimization Initiative in

Bangalore. She is on the editorial team of the *Journal of Industrial Ecology* and an Adjunct Fellow at the Ashoka Trust for Research in Ecology and the Environment (ATREE), India.

Atsushi Watabe is a researcher at the Institute for Global Environmental Strategies (IGES). He has worked for international cooperation projects and programmes addressing climate change mitigation and adaptation in Asian countries, and post-disaster recovery of Japanese communities. His current research focuses on the activities and initiatives for sustainable usages of resources and energy at the household and community levels, for several cities in Southeast Asia.

Dunfu Zhang is Professor at the School of Sociology and Political Sciences, Shanghai University. His publications cover the sociology of consumption, consumer culture and social change in China. His visiting professorships include Harvard University and the University of Manchester, where he worked with colleagues with common interests.

Introduction

*Marlyne Sahakian, Czarina Saloma
and Suren Erkman*

What could be more ubiquitous than eating a meal? Ingesting food is essential to human life and is practised by people across the planet – including the many millions who lack adequate access to food. Yet what and how we eat is part of complex systems, which involve not only food production, but also the cultural, economic, environmental and social aspects of food production and consumption. While eating food is a local activity, it relates to both local and global food chains, involving resource inputs at different stages and outputs with impacts at different scales, from local pollutants to global climate change. The consumption of animal products has tripled in developing countries in the last decades, in addition to rising health concerns in relation to the growing consumption of processed foods (Myers and Kent 2004; FAO 2004; Kearney 2010). These shifts in diet entail massive changes in land and energy requirements for food provisioning, with related environmental impacts (Duchin 2005). Food waste is also a growing issue, with an estimated 30 per cent of edible food wasted globally in 2010 according to the Food and Agricultural Association (FAO) (Gustavsson et al. 2011). Although food losses occur across the system, from food production to storage and transport, food waste at the level of households in developing countries is a growing area of concern (Parfitt et al. 2010; Gustavsson et al. 2011).

Eating a meal is a biophysical activity that draws from and affects ecosystem services, but it is also very much a social and cultural affair. Food cultures in different contexts relate to habits, rituals, preferences and tastes when it comes to preparing, eating and sharing a meal. These can hardly be considered static, as local and global flows of food, people and ideas also influence what food is available in different contexts, how it is stored and prepared, and what is considered tasty or revolting. The daily practices related to preparing and eating a meal unveil how food consumption continues to evolve in different food consumption spaces, when eating at home or when dining out. Urbanization, increased affluence and market liberalization are typically singled out as the main factors contributing to changing food consumption patterns, yet these changes are usually understood as involving an understanding of consumers as individuals, acting rationally in relation to a changing marketplace. The dynamic relation between people and food provisioning systems is far more complex, revealing an interaction between social norms, institutional context, time pressures, everyday routines, competencies, the material dimension of consumption and how all of these evolve over time and across spaces. This perspective has led to a growing body of literature in recent years that seeks to understand eating as a social practice (Plessz et al. 2014; Warde 2013; Halkier and Jensen 2011), and food waste as embedded in social and material contexts (Evans 2012).

The aim of this book is to uncover how food consumption has been changing in Asia and the Pacific in the past decades, and in what way these changes can be explained through an understanding of food consumption and related waste at the intersection of social practices and biophysical considerations, including environmental impacts. Nowhere are changes in food consumption patterns and practices more observable than in urban areas. In this volume, Hellmuth Lange refers to cities as 'breeding grounds of change'. Cities in Asia and the Pacific are sites of food consumption for millions of people, and offer a wider choice of foods and greater exposure to foreign food than in rural areas. They are also sites of shifting domestic relations, with more women in the workforce, which does not necessarily entail more men in the kitchen, but can place a premium on convenience when it comes to food preparation and consumption. Among households in urban settings across the region, the notion of preparing and eating a meal ranges from opening a can of processed meat or adding hot water to dried noodles, to elaborate home-cooking, conspicuous buffet meals in restaurants or fast food meals in air-conditioned shopping malls. Even within the same city, food consumption practices and patterns vary considerably between differing groups of people, and practices can change in different spaces of consumption from food preferences when eating at home, to eating out.

This book brings together researchers from environmental studies, anthropology, sociology, tourism, architecture and development studies, along with practitioners working in the area of food consumption, waste management and the policy arena, to provide a glimpse into food consumption trends in several cities in Asia and the Pacific. The emphasis is downstream from production processes, with a focus on consumer-citizens, illustrating food consumption at home and when eating out, and related food waste. This interdisciplinary volume relates different scales of analysis, and combines different methodologies and approaches, from empirical research among households, to the analysis of national data and regional trends. The contributions include perspectives from China and India, as well as research on several cities across the region including, in the order in which they are presented: Ho Chi Minh City, Vietnam; Metro Manila, the Philippines; Honolulu, Hawai'i, USA; Melbourne, Australia; Bengaluru, India; Tokyo, Japan; Shanghai, China; Kuala Lumpur, Malaysia; and Seoul, South Korea.

Apprehending (un)sustainable food consumption

The interdisciplinary nature of this book compels us to describe some of the main approaches our authors engage with in understanding food consumption in its multiple dimensions. What follows is a brief introduction to two main conceptual handles, which guided the development of this volume: food in its biophysical dimension in relation to environmental impacts, and food as a cultural activity and part of social practices.

Food patterns and environmental impacts: the biophysical dimension

Food production and consumption can be apprehended as a biophysical process (a metabolic activity *par excellence*), involving material and energetic provisioning from resources that draw from and depend on ecosystem services. Environmentally sustainable food production and consumption therefore implies minimizing negative environmental impacts, such as resource depletion, loss of biodiversity or pollution, and reducing excessive throughputs. As such, food production and consumption is submitted to the constraints of the biosphere, as put forward in the work on bio-economics by Georgescu-Roegen (1966, 1971, 2006), and further developed in ecological economics (Boulding 1966; Daly 1968). Industrial ecology is a related field of practice and research, which offers methodological tools for the efficient management of resource stocks and flows within the conceptual framework of scientific ecology. As put forward by Frosch and Gallopoulos (1989), the use of the term 'industrial ecology' suggests that industrial systems could draw inspiration from natural systems – including household activities in an industrialized society (as 'industrial' here has a broad meaning, referring to all activities taking place within the industrial society). It would follow that such industrial systems emulate the operational efficiencies found in nature (Erkman 1997). This entails assessing the industrial metabolism of a biophysical activity, evaluating the quality and quantity of energy and material stocks and flows in a given system, and understanding how that system could be designed to perform with less environmental impacts at different stages – from the extraction of natural resources, through production, consumption and final disposal (Ayres and Simonis 1994). This perspective broke with earlier understandings of environmental issues, which focused mainly on cleaning up 'end of pipe' waste and pollution, and progressively developed, starting from the 1960s. Looking beyond the end of pipe approach in an industrial ecology and ecological economics perspective, the complete flow of materials and energy is taken into account, including how food is produced, stored and distributed, through to consumption and final waste disposal.

Early work in industrial ecology focused primarily on production processes: industrial parks and factories in the developed world were the focus of enquiry, with some pioneering work in the context of developing countries (Binder 1996; Erkman and Ramaswamy 2003). A 2005 special issue of the *Journal of*

Industrial Ecology broke with this trend by placing an emphasis on consumption: the editors underlined the fact that '[m]ore work is needed that connects household metabolism research with a deeper understanding of social mechanisms and consumption patterns' (Hertwich 2005: 5). Two main methodologies are relevant to understanding the biophysical dimension of food consumption: Material Flow Analysis (MFA) and Life Cycle Assessment (LCA). MFA approaches seek to assess the material and energy flows and stocks related to activities, such as household food consumption, and to then quantify the material and energy flows within this system (Brunner and Rechberger 2004). Rather than focus on activities, LCAs consider the environmental impacts of products and services across their life cycles, from extraction through production, consumption and final waste (Jungbluth 2000; Jungbluth et al. 2000; Guinée 2002; Jungbluth 2011).

In this volume, several authors consider the biophysical dimension of food consumption, relating the environmental impacts associated with types of food, such as meat (Chapters 1 and 7). The activity of household food consumption can be assessed through an MFA, while the impacts of different products and services can be evaluated through a Life Cycle Assessment. Chapter 2 engages with LCA literature in relation to agriculture and food waste management. In Part III, two contributions engage with MFAs to calculate food waste in a restaurant (Chapter 10) and a school cafeteria (Chapter 12).

Food meanings and practices: cultural and social dimensions

Social practice theories are based on understanding consumption and production as involving a dynamic relation between people, social context and a material world. Attention is deflected away from individual and rational choice models (see Shove 2010 for a critique). Instead, it is the routine and habitual practices of everyday life that become the object of study. Starting in the 1980s, consumption studies had a more cultural reading: consumption was seen as being made up of rituals, with goods coded with meaning, as symbols that demarcate status (Douglas and Isherwood 1979; Baudrillard 1968; Appadurai 1986; McCracken 1988), with a vast amount of work on material culture and consumption (Miller 1995, 2001a, 2001b). While some forms of food consumption can be considered 'conspicuous', such as elaborate buffets at weddings, much everyday food consumption is highly habitual and inconspicuous (Shove and Warde 1998). For Warde, eating is a particularly interesting area of investigation because it brings together 'the material, the corporeal and the mundane, and by repetition, routine and convention' (2014: 288). Eating is also a precondition for passing judgement on taste: the practice precedes the judgement (ibid).

There are several interpretations of what makes up a practice (Schatzki 1996; Reckwitz 2002; Shove and Pantzar 2005), building on the groundbreaking work of Bourdieu (1977) and Giddens (1984) who overcame the agency-structure debate in social sciences by conceptualizing structures as both medium and outcome of the practices they recursively organize. One perspective is that social practices are a 'block' of routinized actions, (re)produced and ordered in space and time, through the distributed agency across three elements that make up a practice: people, their competences and understandings; the material dimension, including objects, infrastructures and technologies; and the social context, involving culture, norms and institutions. Research that relates social practices to food consumption has been undertaken, albeit mostly in the context of Europe (Halkier and Jensen 2011; Warde 2013; Jaeger-Erben and Offenberger 2014; Sahakian and Wilhite 2014). Approaches that combine social practice with a biophysical understanding of food consumption are less common, with an emphasis on this approach in the Industrial Ecology Group at the University of Lausanne, involving two editors of this volume (Erkman and Sahakian; see also Burger Chakraborty et al. 2016; Leray et al. in press). One important aspect of social practices is that they can change over time: how we go about our daily lives keeps evolving, across different contexts. Refrigerators may have become a norm for certain people living in urban centres in Asia and the Pacific, but remain a luxury for many others; they are associated with convenience for some, class status for others who covet large-format refrigerators (popular in North America), and unhealthy eating habits for others who find that refrigeration deteriorates the nutritional value of food. Wilk (2006) describes the incredible variety and heterogeneity of the food system, 'its protean creativity and unpredictable trends' (2006: 13), while underlining the tensions with the industrial food systems, concerned with quantity, convenience and profit. The tensions and creativity in relation to how food consumption is changing across the region are very much revealed by the different accounts offered in this volume.

Some consider the interplay between the competencies of certain groups, on the one hand, and collective conventions that often involve confrontations with social boundaries (Chapter 3, on gender and generations; Chapter 7, on caste), on the other. Others focus on social and material contexts at the intersection of biography and history (Chapter 3, 8 and 9 on the experience of wars and famine; Chapter 5 and 6 on migration and cultural flows), as issues of space and the built environment (Chapter 4), and of infrastructure (Chapter 6 on policy and regulation; Chapters 8 and 11, on systems of provision). Regardless of emphasis, all of the accounts engage with local contexts, and how these relate to trends in food consumption and waste generation. In particular, Chapters 3, 7, 8 and 9 offer reflections on what these mean in an unequal city. In some cases, people are considered in terms of socio-economic groups, such as the middle classes, while other are organized in terms of cultural patterns and social practices, such as vegetarians. This raises questions around how to frame the object of analyses, recognizing social diversity while not foregoing notions of class and the power structures implied.

Organization of the book

The book is organized into three parts: Part I considers regional patterns and trends, as well as the environmental impacts associated with different food production and waste management systems – in varying contexts in Asia and the Pacific. In Part II, food practices and cultures are discussed, including the social meanings of eating at home, eating out, eating local and eating global. Part III examines food waste dynamics in Asia and the Pacific, at both the level of households and the service sector. Here, the metabolism of food consumption and related waste is considered, in relation to social practices and the biophysical dimension of food.

In Part I, the first chapter by Hellmuth Lange raises awareness around the differing levels and patterns of meat consumption in developing countries. With a focus on China and India, he presents a more nuanced picture of these trends using two analytical perspectives: a) a vertical perspective by comparing different income levels among the new middle classes and how this relates to meat consumption, and b) a horizontal perspective, reflecting different professional, educational and cultural 'milieus' and lifestyles co-existing within a particular income level – comprising different and even contrary orientations and patterns of meat consumption and ecological concerns. He demonstrates that even on the upper income levels of the new consumers, meat consumption on a per capita basis in these places is still significantly lagging behind the Global North and will continue to do so. This holds all the more true for the middle and lower income groups of the new consumers, which constitute the vast majority of the 'new middle classes'. However, even small increases in meat and aquatic product consumption will dramatically increase ecological footprints due to the large number of people involved.

Megha Shenoy adds to the first section of this book with an environmental assessment of sustainable food production and consumption, covering several cities in Asia, in Chapter 2. Drawing on industrial ecology methods, Shenoy presents Life Cycle Assessment (LCA) as a valuable tool to identify strategies that could result in lower environmental impacts. With most countries lacking adequate LCA inventory data, the strength of her contribution is to go beyond the challenge presented by limited availability of data, to showcase how life-cycle thinking can be brought to bear on sustainable food consumption issues. Using a combination of a literature review of LCA studies and an examination of local practices and policies, Shenoy provides an overview of the diverse ways in which food production and waste are being handled in four cities: Bengaluru, Kuala Lumpur, Metro Manila and Seoul. The setting in which food production and consumption takes place is of significance: scenarios can be envisioned based on life-cycle thinking, but solutions are highly context dependent and not necessarily transferable to other cities. She then offers recommendations for lowering the environmental impact of agriculture by comparing organic agriculture over non-organic practices, and considers low-impact options for treating food waste segregated at source, including bio-gasification, wet and dry animal feed production, and compost production over incineration and landfills.

In Part II, several contributors engage with the social and cultural dimensions of consumption.

In Chapter 3, Judith Ehlert delves into Ho Chi Minh City's growing eating out scene, which is fuelled by Vietnam's rapid global economic integration, food system modernization and re-emerging urban middle classes. Drawing on data collected during long-term field research, Ehlert approaches food consumerism from the emic perspective of 'new consumers' and discusses food sharing as a performance of 'middle-classness'. She shows how the gendered and generational dimensions of identity construction, social belonging and boundary drawing are locally embedded in food practices as well as in broader societal changes. Vietnam's development has not, however, led to the embodiment of 'modern', globalized food consumerism alone: Ehlert argues that social inequality is also evident in those not benefiting from the overall economic boom, eating or recycling others' food waste, thereby symbolizing the ingestion of poverty and an 'inferior' identity. Nascent initiatives embedding global consumerism critiques as local 'counter-culture' are beginning to tackle this imbalance.

In Chapter 4, Czarina Saloma and Erik Akpedonu explore condominium living as a visible marker of a changing urban landscape and contemporary culture in Metro Manila, a mega city of twelve million people and the capital region of the Philippines. Whereas houses in other residential arrangements form neighbourhoods that are spread across, those in the condominium form a neighbourhood that is spread upwards; hence their term, 'vertical neighbourhood'. Using key informant interviews with condominium residents and a practice approach, the authors highlight the significance of the material dimension of food consumption and the dynamic relation between different elements of a practice. They show that the vertical neighbourhood places urban residents in a new material setting, expanding social patterns and practices of food consumption. Condominium design typically reconfigures the home, hitherto the setting of the preparation of food and its enjoyment, by moving cooking and eating practices out of the residential unit. However, new routines emerge as residents innovate on familiar ways of consuming food at home. The chapter ends with an account of the consequences of food consumption practices in condominiums on waste generation and energy consumption, and their implications on the creation of a liveable and viable mega city.

Elizabeth Louis and Kyle Datta delve into the notion of 'local food' in Chapter 5, describing how this contested notion has steadily become a metaphor for sustainability in Hawai'i, understood by some as healthy, safe, just and environmentally sound food. Using data collected on consumer buying habits in Honolulu in 2009, the authors discuss different consumer segment preferences for local food in Honolulu, and trace the differential histories and social contexts that have given rise to varied purchasing patterns. Select staple goods were chosen for the study, involving beef, eggs, milk, lettuce, tomatoes and bananas. The consumer preference model found that price parity, access and the ability to identify the product as local were the three most important attributes that defined customer choice across commodities and ethnic groups. In addition to these, they suggest that historical and spatialized food identities influence consumers choices for local food in Hawai'i. For some, products specifically grown in Hawai'i take on a value beyond the economic or environmental benefits, to become a social good, imbued with a sense of pride and ownership. A gap persists, however, between desire for local food and its consumption. To further understand this issue, the interface between consumers and producers is taken into account, with a consideration for the main barriers towards increased 'local' food production in this context.

In Chapter 6, Warwick Frost and Jennifer Laing take us to the 'foodie city' of Melbourne, where changes in food consumption patterns, cultures and cuisines have been underway over the last three and a half decades. During this period, both domestic food consumption and the tourist experience have transformed in response to new dining opportunities outside the home. Frost and Laing demonstrate how distinct features of café and restaurant culture in Melbourne, such as al fresco dining and the prominence of small, independent operators, have been bolstered by a favourable policy and regulatory framework. The growth of diaspora cuisines and their spread to the more general population is given particular attention, with a focus on the Chinese and Italian diasporas. The notion of healthy eating is highlighted in relation to eating at home, where recent studies among Melbourne households raise concern over increases in fast food consumption and decreases in vegetable intake in daily diets. This contribution thus reveals a paradox between the varied and thriving restaurant culture, and poor-quality and high-calorie home consumption. The authors, likewise, emphasize environmental concerns in relation to these food consumption trends, including the tension between population growth, urban sprawl and local food production, and what these entail in terms of rising food prices and increased food mileage.

The practices of Indian vegetarianism are discussed in relation to limited resources in Chapter 7, with a focus on the information technology capital of India, Bengaluru. Rune-Christoffer Dragsdahl points out that India is often described as the exception from the global pattern, whereby increased economic growth leads to increased levels of meat consumption, with per capita meat consumption at solely 5 kg per year (FAO 2009). Without comparison, India therefore also represents the country in the world with the largest potential future increase in meat consumption, which would have significant impacts on food resources and environmental sustainability as outlined by Hellmuth Lange in Chapter 1. Based on ethnographic material gathered among youth in Bengaluru between 2009 and 2010, this chapter explores the practices of vegetarianism and non-vegetarianism. As caste structures gradually weaken over time, meat consumption could rise significantly, yet the veg–non-veg divide is already commonly challenged when practised. The ambiguity of meat eating among lower castes is emphasized, as well as the importance of egalitarian friendships and peer pressure in making young people from vegetarian families start eating meat. Dragsdahl argues that youth in Bengaluru practise balancing of, on the one hand, influences from their family and caste, and on the other hand, influences from their friends and romantic relationships. His work points to the blurred lines between vegetarian and non-vegetarian food items, and implications for sustainable food consumption pathways.

Part III focuses on food waste dynamics, a key theme in sustainable consumption studies. The section starts with a contribution from Japan, by Atsushi Watabe, Chen Liu and Magnus Bengtsson (Chapter 8). The authors consider the question of uneaten food in relation to changing food consumption practices in Japanese society over time, embedded in specific cultural and social contexts, but also influenced by changing systems of provision. Food waste estimates are discussed, at the household level, as well as the numerous food-related practices that have led to what the authors call a 'distancing in social relations'. In addition to the distancing of consumers from supply chains, this form of social distancing creates less opportunities for people to engage in practices to reduce food waste, as had been prevalent in the past. The authors argue that solutions to food waste issues require a broad perspective, which they term 'social responses'. Three examples are highlighted in particular: 'local loops' which connect farmers to consumers; food banks that give consumers an opportunity to share unwanted food; and salvage parties, where participants cook and eat food together. These innovative examples address the issue of bringing people together to learn new practices by reducing social distance.

Turning now to Shanghai, in Chapter 9, Dunfu Zhang studies food consumption and related waste in the mega city Shanghai, against the backdrop of a country that boasts the world's largest population, a long agricultural history, limited natural resources, increasing environmental pollution and the collective memory of severe famines. Engaging with social practice theory as a theoretical framework and drawing from informal interviews in Shanghai, Zhang finds that most people do not like to waste food, particularly people with experiences or memories of food shortages. Making use of leftovers is tied to moral issues in this respect, but also caring for family members and other relations of intimacy. Sharing leftovers is a demonstration of loving relations. Among a younger population, there is also a growing discourse around environmental concerns related to food waste. Zhang engages with the moral dimension of food waste, drawing attention to the distinction between frugality and thrift towards sustainable pathways. He argues that, over time, health aspects and notions of altruism, such as caring for family and for others, could be further reinforced. Through these two themes, people in Shanghai and possibly in other urban contexts in China could transition from being thrifty consumers to consumer-citizens, towards more sustainable lifestyles.

In Chapter 10, Effie Papargyropoulou illustrates the question of food waste in Kuala Lumpur, Malaysia, through the case of an upscale hotel restaurant. The case study connects the social practices and biophysical patterns of food waste through an interdisciplinary research approach, drawing from ethnography, industrial ecology and grounded theory. The case study uncovers a diverse range of drivers that give rise to food waste; some linked to social norms and behaviours of the consumer, some associated with the organizational structures, operations and policies of the restaurant, and others related to the type of food service provided, menu design and food preparation method. The findings suggest that multi- and cross-level interventions are required to achieve food waste prevention. Most importantly, the study argues that food waste should be studied as part of food consumption rather than as a separate issue, and as such recognizes it as inherently related to social practice and not only the material context within which it is generated.

In Chapter 11, Keith Lee explores food waste in Seoul's households at the interface of changing food retail systems, food practices and systems of provision. Based on interviews with Korean households and observations made in Seoul's diverse food retailers, Lee demonstrates the reflexive relation between changing everyday life and the retail landscape – resulting in a possible exacerbation of food waste. Growing time-scarcity and shrinking household size have led to more convenience in how food can be purchased and prepared, in addition to growing popularity around the notion of grocery shopping for leisure. As Lee points out, these trends may lead to an increase in household food waste if consumers do not match this increased access to food with the appropriate food management competencies. More broadly, the chapter illustrates the importance of accounting for systems of provision in studies of household food waste, as well as the relation between social practices and systems of provision.

Chapter 12 shifts attention to the meso-level of food consumption: school and university canteens. Considering the case of a private school in Metro Manila, Abigail Marie T. Favis and Rafael Deo F. Estanislao consider food flows in their biophysical dimension, in addition to contextualizing food flows in specific social and institutional settings. They begin by sharing results of an audit conducted on rice plate waste at the primary school and the university level. In addition to uncovering quantities of wasted rice, the authors seek to understand why such waste occurs, engaging with a social practice approach. The interplay between the material dimension of food consumption in a school cafeteria, routinized bodily practices of students and canteen staff, as well as the social context of rice waste generation all come together to tell a comprehensive story of why waste is generated, and what measures might be put in place to reduce rice waste in this setting.

Food consumption and waste: trends across the region

The contributors to this volume were given the challenge of approaching food consumption in its environmental and social dimensions, and from the perspective of households. One essential finding is the significance of local contexts in understanding food consumption patterns, and how these contexts and associated practices evolved over time. Differences exist between countries, as can be expected, but variations also exist between different groups within cities. Research on food consumption may involve the mundane activity of ingesting a meal, but reflects an endless variety of interpretations – from how ingredients are purchased and prepared, to how they are consumed, shared and wasted. Several trends emerge from reading these contributions from urban Asia and the Pacific, which recall early work on social practices by Bourdieu (1977) and his emphasis on class distinctions in relation to food consumption in France. Socio-economic class distinctions are evident, across the chapters. In Chapter 1, Hellmuth Lange points out that both income levels and differing professional, educational and cultural 'milieus' and lifestyles are relevant to understanding varying meat consumption patterns in China and India. Echoing this sentiment and in Chapter 7, Rune-Christoffer Dragsdahl's empirical research among middle-class youth in Bengaluru (India) demonstrates 'the ambiguity of eggs, selective vegetarianism, and contextual meat consumption', where meanings around vegetarian and non-vegetarian meals relate not only to caste, but also class. In their analysis of recent surveys on food consumption among Melbourne households, Warwick Frost and Jennifer Laing (Chapter 6) underline the relation between socio-economic status and position, in relation to food purchase and consumption habits. In relation to organic food consumption, Megha Shenoy (Chapter 2) also notes that organic produce is more affordable to the upper-middle classes and elite populations, than it is to the general population. In places like Metro Manila, local organic production is promoted, but is mostly destined for export to wealthier consumers elsewhere. For Judith Ehlert (Chapter 3), however, and in the case of changing food consumption practices and the eating out trend in Ho Chi Minh City (HCMC), Vietnam, class cannot be seen as a fixed disposition in which food consumption practices evolve but as something in flux. Other dispositions are equally relevant, related to social norms around inter-generational and gendered practices of food sharing. The notion of social inequality in relation to food consumption is a relevant theme that Ehlert explores in her work, whereby food over-ordering may be a social marker for doing middle-classness, and at the same time ingesting recycled food (wasted by someone else) clearly symbolizes the ingestion of an inferior identity and social status. As she puts it, 'In a context of economic growth that is not equally distributed but has led to new forms of poverty including food scarcity, the food practices of eating out and over-ordering come to represent the growing inequality and exclusion of those not benefiting from the economic developments.' The theme of social inequality in relation to food consumption would merit further exploration, as inequalities persist throughout the region.

The historical significance of food consumption is also apparent in nearly all the contributions, or how preferences, habits and rituals have evolved over time in different contexts. The historical dimension relates to how 'local' food is interpreted in Hawai'i (Chapter 5) in the work of Elizabeth Louis and Kyle Datta. The influence of migratory flows on local food consumption practices is revealed – both in relation to food preferences in Honolulu, and in relation to eating out and the burgeoning of a vibrant restaurant and café scene in

Melbourne (Chapter 6). In Chapter 8 by Atsushi Watabe, Chen Liu and Magnus Bengtsson, the authors demonstrate how food consumption practices have changed in Japanese society over time, embedded in specific cultural and social contexts, but also influenced by changing systems of provision, including the retail and restaurant environment. Another element cuts across the different sites of food consumption in urban Asia and the Pacific is that of time pressures and a quest for convenience when it comes to food consumption. In relation to an industrial agrarian system that seeks efficiencies in (re-)production processes, certain households turn to convenient food – as is the case among HCMC households in the work of Ehlert (Chapter 3). The relation between production, retail and consumption is emphasized in Keith Lee's contribution (Chapter 11), recognizing a dynamic relation between a quest for convenience, the changing retail landscape and household food waste in Seoul. As he succinctly states, food consumption is also about coordinating practices, yet '[c]oordinating practices can be tricky considering they take up time, a finite resource'. Time as a resource, in addition to material and energy, is a highly relevant area of further study across urban Asia.

Eating out emerges as a key theme in several chapters. As Hellmuth Lange points out in Chapter 1, eating out is less related to the 'capitulation to Western patterns of taste or the business model of "McDonaldization" described by Ritzer (2011), and more to its suitability for urban customers under the conditions of today's urban reality'. Eating out, pre-cooked meals and convenience food are quite fitting to urban constraints and opportunities, he argues. Eating out becomes a key site for redefining how meals are shared, why food is wasted, and how gendered and inter-generational dimensions of food consumption are negotiated in HCMC (Vietnam), as described in Chapter 3 by Judith Ehlert. The vertical neighbourhoods of Metro Manila (Chapter 4), often built in unison with shopping malls that feature a variety of restaurants, illuminate some of these constraints and opportunities. In a sample of households living in all types of residential arrangements, Saloma and Akpedonu note that households in condominiums eat out more frequently than non-condominium households. In Melbourne (Chapter 6), eating out is a site where food consumption patterns and practices have changed dramatically in the past few decades, what the authors Warwick Frost and Jennifer Laing have termed a food revolution. The vibrant and varied restaurant culture, partly due to supportive government regulation and the legacy of waves of migration, exists in contrast with food consumption practices in the home. Rune-Christoffer Dragsdahl (Chapter 7) also demonstrates how food consumption practices change when people step out of the boundaries of the home: the restaurant offers a different context for food consumption that breaks with the family sphere and where peer groups renegotiate food preferences.

Rather than focus solely on household food waste, eating out also becomes a key site for tackling this growing issue. Effie Papargyropoulou (Chapter 10), and Abigail Marie T. Favis and Rafael Deo F. Estanislao (Chapter 12) consider the case of restaurants in Kuala Lumpur and school canteens in Metro Manila, respectively, raising the issue of food waste in the service industry. Practices outside of the home, such as school canteens, could positively influence food waste reduction in the home, as suggested by Favis and Estanislao. Spaces of consumption bring in the broader theme of the material dimension of consumption, including infrastructure, objects and technologies. The role of the material world in relation to social practices is relevant, in terms of how food is presented, when and how it is made available and in what settings. In both the examples of food consumption in Kuala Lumpur high-end restaurants and Metro Manila private school canteens, the material dimension of food consumption is also directly related to food waste: in how menus are designed and buffets are presented, for the former; and in how rice is presented and served, for the latter. The contribution by Czarina Saloma and Erik Akpedonu (Chapter 4) also illustrates how 'vertical neighbourhoods' made up of condominiums place limits on how food is prepared and enjoyed in the home, due to space constraints, the high cost of electricity and regulations prohibiting in-home entertainment. In Judith Ehlert's work on eating out in HCMC (Chapter 3), the individualization of eating is supported by Western-style meals, where individual plates are more common than shared dishes. She notes how this style of eating may lead to renegotiating gendered roles around commensality, or caring for how others share in a meal.

Tensions in how food consumption is practised in urban Asia and the Pacific are also apparent in this volume. Across several cities in the region, there are tensions between existing regulations when it comes to food waste handling, and their proper enforcement – or a gap between policies and practices, as detailed in Chapter 2 by Megha Shenoy. In Hawai'i (Chapter 5) and in a study of meanings around 'local' foods, the authors find that a gap persists between a desire for local foods, and the actual practice of purchasing and consuming local foods. Willingness to pay is one factor to explain this gap, but in addition to the affordability of products, secondary factors such as the availability of foods and knowledge of local food options, and cultural preferences for certain foods, are also considered. In the Melbourne example (Chapter 6), a gap persists in the quality and diversity of food consumption practices at home and when eating out: while the restaurant scene presents a sophisticated and varied food culture, eating at home in Melbourne entails fast food consumption, and limited fruits and vegetables in everyday diets. In the case of Shanghai, the tensions are related to how food waste is handled among households, as detailed in the work of Dunfu Zhang (Chapter 9). For some people, making use of leftovers is tied to moral issues, as well as caring for family members; for a younger population, food waste may be more related to environmental concerns. Zhang discusses how the notions of frugality and thrift can be differentiated, and what this entails for food waste in the future. Judith Ehlert (Chapter 3) notes that younger generations are not only embedding global consumer culture in their everyday practices but, at the same time, connecting to global consumer criticism and its concern for environmental sustainability and growing social inequity, as inherent in the food system.

Several contributions are optimistic about what the future might hold in terms of food consumption patterns and related environmental impacts. In Chapter 2, Megha Shenoy outlines initiatives underway in four Asian cities: Bengaluru, Kuala Lumpur, Metro Manila and Seoul. She considers how agricultural practices and food waste treatment is evolving in different contexts, with recommendations in relation to life-cycle considerations. In general, she recommends lowering environmental impacts by favouring organic agriculture over non-organic practices, reducing the amount of food waste generated, and treating food waste that is segregated at source by bio-gasification, wet and dry animal feed production, and compost production over incineration and landfills. However, the specific regulations to translate these methods into practice may have to be tailored for each city, keeping existing local conditions in mind. Abigail Marie T. Favis and Rafael Deo F. Estanislao (Chapter 12) give some examples on how changes in school canteens were effective in reducing food waste, including the removal of free-serving rice bowls from tables in primary schools, and efforts to create a new norm around the undesirability of wasting food. These forms of social learning are also evident in the examples of what the authors term 'social responses' to the issue of food waste in Japan (Chapter 8). The three examples they provide point to the limits of information campaigns and new opportunities presented by social interfaces around food and related waste, at a societal level, rather than an individual or household level. Practices such as salvage parties and food banks are opportunities for new relations between people in Japanese cities, involving new narratives around reducing food waste.

Sustainable food consumption: towards inter- and trans-disciplinarity

This edited book aims towards an inter-disciplinary reading of changing food consumption practices and patterns in Asia and the Pacific. Contributors sought to combine a biophysical understanding of consumption with the view of consumption as tied up with everyday social life in specific contexts. In most cases, they explore research questions from the vantage point of their disciplines then borrow from other disciplines to reach beyond disciplinary boundaries. These approaches often bring to light the fundamental differences in how researchers frame problems, and how disciplinary backgrounds influence the definition of the research object and goals of scientific inquiry (Fahy and Rau 2013; Burawoy 2013). For some, environmental impacts are the setting against which the object of study gains in relevance: first, a particular pattern of consumption is considered to be significant in terms of the using up of environmental resources and related impacts (such as meat consumption or food waste), then the pattern is described and defined in terms of social practices and/or cultural meaning. For others, just the opposite takes place: new practices of food consumption are described, and then related to environmental considerations. Both are valid approaches towards interdisciplinarity. As Defila and Di Giulio point out (2015), the objective of inter- and trans-disciplinary approaches can either be to advance theoretical and empirical knowledge, or to contribute to solving problems and improving practices. Through these contributions and interdisciplinary reflections, authors seek to describe problems if not solve them – in the case of food waste for example – and to improve practices, against the backdrop of environmental considerations.

The notion of trans-disciplinarity is more difficult to apprehend and achieve: it involves cutting across or transcending disciplinary frameworks, as in the metaphor of a trans-Atlantic journey across continents that aims towards reaching new intellectual spaces. Assessing both 'how much' materials and energy a household consumes in preparing and eating a meal, as well as 'why' and 'in what way' those consumption patterns persist or change requires a considerable amount of resources and skills on the part of the researcher. The contributions of Effie Papargyropoulou in Malaysia (Chapter 10), and Abigail Marie T. Favis and Rafael Deo F. Estanislao in the Philippines (Chapter 12) demonstrate the amount of work involved in this kind of endeavour: measuring food waste is a time-consuming and resource-intensive task, which explains why so little is known about food waste quantities around the world and in different contexts, beyond memory recall methods. Their work goes further, as they demonstrate the many elements that come together to create food waste. Some of these elements relate to social norms and policies, while others relate to how food is presented on menus, prepared and served. As discussed elsewhere and in relation to food waste in Bangalore (Leray et al. in press), reflexivity between concepts and methods implies engaging in an iterative process, going between and beyond disciplinary approaches. In addition, several contributors show the value of historical perspectives, which also raise empirical research challenges: how can both qualitative and quantitative research be garnered over time, to then analyse changes underway? Memory recall methods are one approach, but contributors in this edited book generally favoured in-depth interviews and observations, which can be difficult to undertake over an extended period of time. The time and resource investment in these qualitative and ethnographic methods has a clear added value, however, as Effie Papargyropoulou points out in her chapter: 'It is this intensiveness, however, that provides the depth and richness in data analysis required to understand food waste generation.'

In terms of resolving practical questions about increased resource consumption or how practices are changing over time, questions related to time and space become significant: in this respect, we can see time and space as cutting across social practices. Assessing household food consumption practices within the bounded frame of a 'home' is one thing, but very quickly the retail–consumer interface, or the influences of the world of work on household food consumption become apparent. This raises empirical questions in how household food consumption can be studied and what boundaries need to be set, if any, between consumption and production. One concept that bridges environmental and social sciences in relation to food consumption is that of 'systems of provision' (SOP) – first developed by Fine and Leopold (1993). The SOP approach combines technological, economic, social and cultural dimensions to how goods and services are distributed, marketed and consumed, from production to consumption. In this perspective, '[c]ommodities are socially constructed not only in their meaning but also in the material practices by which they are produced, distributed and ultimately consumed' (Fine and Leopold 1993: 15). The SOP approach considers each link in the production to consumption chain, and its role in terms of both material and cultural aspects (Fine 2002). Both Keith Lee (Chapter 11), and Atsushi Watabe, Chen Liu and Magnus Bengtsson (Chapter 8) engage with SOP, although arguably many of the contributions look upstream from households to systems of provision. Indeed, much of the existing research that engages with social practices theory has worked at the level of households to uncover daily practices, but without necessarily relating to how practices evolve across the supply chain, in terms of food production for example, or the interface between production and consumption. The role of the retail sector, as well as the restaurant industry, in shaping everyday practices and in turn being shaped by consumer practices would be an interesting area of further study, and one that is explored in several contributions (for example, Keith Lee in the case of Seoul, Chapter 11).

This volume points to the great diversity in how food consumption is practised across the region, and calls for further inter- and trans-disciplinary research – over time – that links micro-levels of household food consumption to meso-levels of consumption, through a systems of provision approach and a consideration for social practices at the consumer–retail interface, and also when eating out, at work or in school. It also points out that often underlying this great diversity of food consumption practices are serious consequences for the environment, as well as social inequality, not only in the distribution of opportunities (such as access to adequate food) but also in the meanings attached to differences (such as consumption as a way to display social status). By locating the discussion of everyday social practices within the biophysical, historical and global patterns of food consumption, this book makes a timely contribution to current debates on consumption in relation to sustainable transformations, and the challenges and opportunities toward socially just and environmentally sound food consumption in urbanizing Asia and the Pacific.

References

Appadurai, A. 1986. *The Social Life of Things: Commodities in Cultural Perspective*. New York: Cambridge Studies in Social & Cultural Anthropology.
Ayres, R. and U. Simonis. 1994. *Industrial Metabolism: Restructuring for Sustainable Development*. Tokyo: United Nations University Press.
Baudrillard, J. 1968. *Le système des objets*. Paris: Gallimard.
Binder, C. 1996. *The Early Recognition of Environmental Impacts of Human Activities in Developing Countries*. Zurich: Federal Institute of Technology Zurich (ETHZ).

Boulding, K. 1966. 'The economics of the coming spaceship Earth.' In H. Jarrett, *Environmental Quality in a Growing Economy*. Baltimore, MD: Resources for the Future/Johns Hopkins University Press.

Bourdieu, P. 1977. *La distinction critique sociale du jugement*. Paris: Les Editions de Minuit.

Brunner, P. and H. Rechberger. 2004. *Practical Handbook of Material Flow Analysis*. Boca Raton, London, New York, Washington, DC: Lewis Publishers, CRC Press.

Burawoy, M. 2013. 'Sociology and interdisciplinarity: The promise and the perils.' *Philippine Sociological Review* 61: 7–20.

Burger Chakraborty, L., Sahakian, M., Rani, U., Shenoy, M. and Erkman, S. 2016. 'Urban food consumption in Metro Manila: Inter-disciplinary approaches towards apprehending practices, patterns and impacts.' *Journal of Industrial Ecology*, 20(3): 559–570.

Daly, H. 1968. 'On Economics as a life science.' *Journal of Political Economy* 76: 392–406.

Defila, R. and A. Di Giulio. 2015. 'Integrating knowledge: Challenges raised by the "Inventory of Synthesis".' *Futures* 65: 123–135.

Douglas, M. and B. Isherwood. 1979. *The World of Goods: Towards an Anthropology of Consumption*. New York: Basic Books.

Duchin, F. 2005. 'Sustainable consumption of food: A framework for analyzing scenarios about changes in diets.' *Journal of Industrial Ecology* 9(1–2): 99–114.

Erkman, S. 1997. 'Industrial ecology: an historical view.' *Journal of Cleaner Production* 5(1–2): 1–10.

Erkman, S. and R. Ramaswamy. 2003. *Applied Industrial Ecology: A New Platform for Planning Sustainable Societies: Focus on Developing Countries with Case Studies from India*. Bangalore, India: Aicra Publishers.

Evans, D. 2012. 'Beyond the throwaway society: Ordinary domestic practice and a sociological approach to household food waste'. *Sociology* 46(1): 41–56.

Fahy, F. and H. Rau (eds). 2013. *Methods of Sustainability Research in the Social Sciences*. London: Sage.

FAO. 2004. *Globalization of Food Systems in Developing Countries: Impact on Food Security and Nutrition*. Food and Nutrition Paper 83. Rome: FAO.

FAO. 2009. *The State of Food and Agriculture 2009: Livestock in the Balance*. Rome: FAO.

Fine, B. 2002. *The World of Consumption: The Cultural and Material Revisited*. London: Routledge.

Fine, B. and E. Leopold. 1993. *The World of Consumption*. London: Routledge.

Frosch, R. and N. Gallopoulos. 1989. 'Strategies for manufacturing.' *Scientific American* 261(3): 144–152.

Georgescu-Roegen, N. 1966. *Analytical Economics: Issues and Problems*. Cambridge, MA: Harvard University Press.

Georgescu-Roegen, N. 1971. *The Entropy Law and the Economic Process*. Cambridge, MA: Harvard University Press.

Georgescu-Roegen, N. 2006. *La Décroissance: Entropie, Ecologie, Economie (1979)*. Paris: Sang de la Terre, Broché.

Giddens, A. 1984. *The Constitution of Society: Outline of the Theory of Structuration*. Berkeley and Los Angeles: University of California Press.

Guinée, J., 2002. *Handbook on Life Cycle Assessment: Operational Guide to the ISO Standards*, p. 692. Dordrecht, The Netherlands: Kluwer Academic.

Guinée, J., M. Gorrée, R. Heijungs, G. Huppes, R. Kleijn, A. de Koning, L. van Oers, A. Wegener Sleeswijk, S. Suh, H. Udo de Haes, H. de Bruijn, R. van Duin and M. Huijbregts. 2002. *Handbook on Life Cycle Assessment. Operational Guide to the ISO*

Standards. I: LCA in Perspective. II A: Guide. II B: Operational Annex. III: Scientific Background. Dordrecht: Kluwer Academic Publishers.

Gustavsson, J., C. Cederberg, U. Sonesson, R. v. Otterdijk and A. Meybeck. 2011. *Global Food Losses and Food Waste: Extent, Causes and Prevention*. Rome: Food and Agriculture Organization of the United Nations.

Halkier, B. and I. Jensen. 2011. 'Methodological challenges in using practice theory in consumption research. Examples from a study on handling nutritional contestations of food consumption.' *Journal of Consumer Culture* 11(1): 101–123.

Hertwich, E. 2005. 'Editorial: Consumption and industrial ecology.' *Journal of Industrial Ecology* 9(1–2): 1–6.

Jaeger-Erben, M. and U. Offenberger. 2014. 'A practice theory approach to sustainable consumption.' *GAIA* 23(S1): 166–174.

Jungbluth, N. 2000. 'Environmental consequences of food consumption: A modular life cycle assessment to evaluate product characteristics.' *The International Journal of Life Cycle Assessment* 5(3): 143–144.

Jungbluth, N. 2011. 'Extending the IOA for investigating the environmental impacts of Swiss consumption and production.' *45th LCA Discussion Forum*. Ittingen.

Jungbluth, N., O. Tietje and R. Scholz. 2000. 'Food purchases: Impacts from the consumers' point of view investigated with a modular LCA.' *The International Journal of Life Cycle Assessment* 5(3): 134–142.

Kearney, J. 2010. 'Food consumption trends and drivers.' *Philosophical Transactions of the Royal Society B* 365: 2793–2807.

Leray, L., M. Sahakian and S. Erkman. 2016. 'Understanding household food metabolism: Relating micro-level material flow analysis to consumption practices.' *Journal of Cleaner Production*, approved, in press.

McCracken, G. 1988. *Culture and Consumption: New Approaches to the Symbolic Character of Consumer Goods and Activities*. Bloomington, IN: Indiana University Press.

Miller, D. 1995. 'Consumption and commodities.' *Annual Review of Anthropology* 24: 141–161.

Miller, D. (ed.). 2001a. *Volume I: Theory and Issues in the Study of Consumption: Critical Concepts in the Social Sciences*. London and New York: Routledge.

Miller, D. (ed.). 2001b. *Volume IV: Theory and Issues in the Study of Consumption: Critical Concepts in the Social Sciences*. London: Routledge.

Myers, N. and J. Kent. 2004. *The New Consumers: The Influence of Affluence on the Environment*. Washington, Covela and London: Island Press.

Parfitt, J., M. Barthel and S. Macnaughton. 2010. 'Food waste within food supply chains: Quantification and potential for change to 2050.' *Philosophical Transactions of the Royal Society B* 365: 3065–3081.

Plessz, M., S. Dubuisson-Quellier, S. Gojard and S. Barrey. 2014. 'How consumption prescriptions affect food practices: Assessing the roles of household resources and life-course events.' *Journal of Consumer Culture*, in press.

Reckwitz, A. 2002. 'Toward a theory of social practices: A development in culturalist theorizing.' *European Journal of Social Theory* 5(2): 243–263.

Sahakian, M. and H. Wilhite. 2014. 'Making practice theory practicable: Towards more sustainable forms of consumption.' *Journal of Consumer Culture* 14(1): 25–44.

Schatzki, T. 1996. *Social Practices: A Wittgensteinian Approach to Human Activity and the Social*. Cambridge: Cambridge University Press.

Shove, E. 2010. 'Beyond the ABC: Climate change policy and theories of social change.' *Environment and Planning A* 42: 1273–1285.

Shove, E. and M. Pantzar. 2005. 'Consumers, producers and practices: Understanding the invention and reinvention of Nordic walking.' *Journal of Consumer Culture* 5(1): 43–64.

Shove, E. and A. Warde. 1998. 'Inconspicuous consumption: The sociology of consumption, lifestyles and the environment.' In R. Dunlap, F. Buttel, P. Dickens and A. Gijswijt (eds), *Sociological Theory and the Environment*. Lanham, MD: Rowman & Littlefield.

Warde, A. 2013. 'What sort of practice is eating?' In E. Shove and N. Spurling, *Sustainable Practices: Social Theory and Climate Change*, pp. 17–30. Oxon, UK and New York: Routledge.

Warde, A. 2014. 'After taste: Culture, consumption and theories of practice.' *Journal of Consumer Culture* 14(3): 279–303.

Wilk, R. (ed.) 2006. *Fast Food / Slow Food: The Cultural Economy of the Global Food System*. Lanham, MD: Altamira Press.

Part I

Trends across the region

1 Same, same – but different

On increasing meat consumption in the Global South

Hellmuth Lange

Consumer research has gained in momentum during the last two decades in its quest for ways to achieve a more sustainable future. Exploring the dynamics of food consumption is one of its most interesting areas of study, with a special issue of *Sustainability: Science, Practice, & Policy* (Sedlacko et al. 2013) providing an instructive overview of the scope and many dimensions of this field. Reisch et al. also indicate a substantial shortcoming: 'although the discussions aim to reflect global trends related to sustainable food, the main geographic focus of both empirical data and policies presented is the European Union (EU)' (2013: 8). Non-sustainable food consumption in the Global South is an issue of sustainability-centred concern often voiced in the Western media by linking criticism of excessive levels of consumption and the new middle classes (hereafter NMCs)[1] as the collective bad guys of consumption. There is also an insufficient yet growing body of research on these groups with a focus on case studies.

The vast number of media reports put forward – more or less explicitly – a disastrous state of affairs: driven by a rage of consumption, the growing middle classes in the Global South are socially and ecologically blind, and are on the verge of undermining all previous and future attempts at advancing a sustainable future, both socially and ecologically. This criticism is reminiscent of 'orientalism' as put forward by Said (1978), yet here the focus is no longer on disparaging the 'foreign culture' in its entirety, but on polarising differentiation within the societies of the Global South. The new middle classes appear as culprits and, above all, the world's poor as victims, as exemplified by the following quote: 'The rising consumer classes in developing countries, especially in emerging market leaders such as China, India, and other Asian countries, will become a serious challenge to global food and resource supplies in the future' and will 'continue to deprive the poor of consumption opportunities' (de Zoysa 2011: 3, 1). Most reporting on trends in meat consumption follows this avenue of thought, with beef portrayed as the tip of the pyramid of non-sustainable nutrition options, a theme that is also picked up by social scientists (cf. among others Dauvergne 2008, 133–168).

In view of the anticipated rise in total meat consumption worldwide, consumption goals should be further considered, along with the suitability of the existing toolkit of information on consumption patterns, incentives and regulations, which have been elaborated thus far by scholars and practitioners in the Global North (see Reisch et al. 2013, Tukker 2008). Before doing so, however, it would be useful to reach a basic understanding on the distribution of, and the emerging preferences in meat consumption – not only between different countries and regions of the world, but also within them – and to consider both quantitative and qualitative (types of meat) parameters. This is the issue dealt with in this chapter, which begins with a comparison between developed and developing countries and differences between developing countries. The examples of India and China (i.e. People's Republic of China) are then used to examine the differences within major developing countries, with a focus on the relevance of regional cultures, differences between urban and rural settings, and the wide array of income brackets. This serves to demonstrate that present patterns of individual meat consumption are still far below the levels of European Union (EU) countries and most Organisation for Economic Cooperation and Development (OECD) countries, and that this will continue to be the case for the foreseeable future. In this respect, Dauvergne's stark warning of a 'global hurricane of consumption' (2008: 4) is misleading. In the final section, the findings on meat consumption in the Global South will be discussed against the background of the bad-guy reputation of the NMCs.

The following two sections build mainly on data provided by the FAO (Food and Agriculture Organisation of the United Nations) and the OECD and, as far as China and India are concerned, on the most recent official National Surveys of both countries (SSB and NSSO, respectively). Differences in scope, accessibility and reliability of data between Chinese and FAO statistics in respect to meat consumption are discussed by Zhou et al. (2012). For the general specifics of statistics on meat consumption, see Hallström and Börjesson (2013).

Meat consumption, patterns and preferences: North–South and South–South

The *global consumption* of meat amounted to 301.7Mt^2 in 2012/2014. An increase of 17 per cent or 51.4Mt is estimated for the next decade, with developing countries accounting for 83 per cent of this increase (OECD/FAO 2015: 136). The average meat *consumption per capita* in kg/year also shows a much biased pattern for developed and developing countries, albeit with a contrasting gradient: the developed countries account for more than twice the amount seen in the developing countries. This gradient is expected to remain stable for the foreseeable future (2024) (Table 1.1).

Comparing developing countries, even more distinct differences become apparent, not least between the prosperous BRIC countries (Figure 1.1).

An even more varied picture emerges when the quantities consumed are broken down into different types of meat (Figure 1.2). The FAO statistics monitor four types of meat: beef (and veal), pork, poultry and sheep. The reference years continue to be the years 2013 and 2023 as the forecast.

Table 1.1 Meat consumption: developed/developing countries, 2012/14–2024

Meat consumption 2012/13–2024 developed/developing countries (per capita kg/year)		
	Average 2012–13	2024 estimated
Developed countries	65	68
BRICS countries	33	35
Developing countries	27	29
World	34	36

Source: OECD/FAO. 2015. *Agricultural Outlook*.

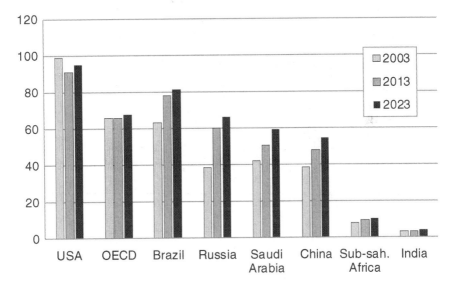

Figure 1.1 Meat consumption, 2003–2023: selected countries; per capita kg/year.
Source: OECD Data 2015, own diagram).

A generally applicable hierarchical order for the preference for different types of meat does not seem to exist, and there are few indications so far that such a structure will evolve in the years ahead. For each type of meat, two or three groups of countries can be distinguished according to the quantities consumed and expected to be consumed during the next decade. The diversity of country-specific profiles also sheds light on the variety of relevant factors.

In terms of per capita consumption, the importance of *beef* is and seems to remain limited, compared to other types of meat. The example of India is not surprising, but there are more countries where this is the case, China included. The fact that *pork* is not eaten in Muslim-oriented countries and Israel is based on a religious taboo which is as important as the taboo on beef in countries shaped by Hinduism, such as India and the like. In countries without such taboos, pork consumption is much more common. China occupies the leading position, but is

26 Hellmuth Lange

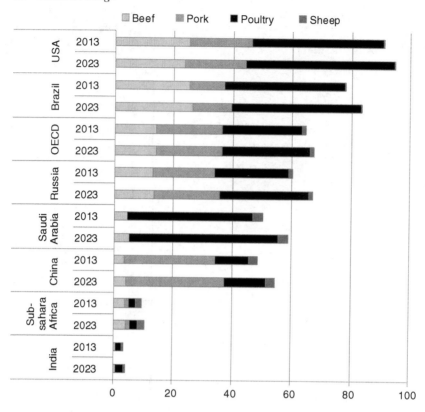

Figure 1.2 Meat consumption, 2013–2023: selected countries/types of meat per capita kg/year.
Source: OECD Data 2015, own diagram.

expected to be outpaced by Viet Nam during the next decade. As for the well-above-average beef consumption in the top group, an interplay of natural and social circumstances in the history of traditional beef production is likely to be of relevance: favourable topological and climatic suitability for large-scale production at relatively low cost, on the one hand, and a long period with a relatively sparse population on the other hand. The economic relevance of beef production in countries such as the United States of America (USA), Brazil, Argentina and Uruguay also relate to the product becoming a symbol of national identity, which includes endowing the act of consuming beef with characteristics of national self-reassurance. This constellation can hardly be exported to other countries.

Compared to other types of meat, *sheep* consumption plays a minor role across most countries while poultry is clearly the frontrunner, and its share is expected to grow significantly during the next decade. In Saudi Arabia, poultry is by far the most popular type of meat. This is also true for the OECD world, although it lags slightly behind here. In the USA, the quantity of poultry consumed exceeds the quantity of beef and pork combined. Even in India, poultry is expected to be consumed more frequently in the years ahead, although still at a low level (1.7kg => 2.2kg). Poultry seems to provide the greatest number of consumption opportunities: it can fill the gap produced by Hindu and Muslim taboos rather than touching a taboo itself; it is easy to raise and its low price also makes it suitable for buyers on low incomes, thus appealing in developing countries as well. Poultry cannot be reduced to an alternative for the poor only, however. It is popular not only across different regions, but also across income levels. Considered to be a symbol of health and sustainability in the meat domain, poultry offers an increasingly attractive option right up to the top of the income pyramid. Since fish also belongs to the full spectrum of animals used for nutritional purposes, data on this issue is also to be considered.

The trend in *fish and seafood* (abbreviated to fish, see OECD/FAO 2014, Note 1) rounds off the options for animal products (except eggs and milk products) destined for human consumption (Figure 1.3). The global production of fish is estimated to be 186Mt in 2023, thus approximately half the expected global meat production, an increase of 17 per cent compared to 2013. No less than 96 per cent of this increase is expected to originate from developing countries, mainly in Asia, where China, India and Indonesia account for 68 per cent of the total growth. Marine and inland aquaculture, still going head to head with marine capture in 2013, will provide the major part of both the total increase and total amount by 2023 (FAO Statistical Yearbook 2013: 146). Trends in per capita consumption in different countries and regions underline the enormous role played by fish in the diet in most parts of the Global South. Africa is falling back, mainly due to further population increases developing

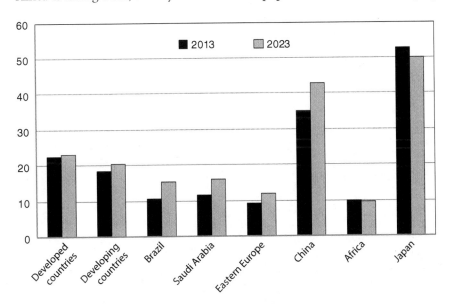

Figure 1.3 Fish consumption, 2013–2023: selected countries; per capita kg/year.
Source: OECD/FAO 2014; own diagram.

faster than the expected supply. China is already leading the field and will continue to do so, with an expected increase of 23 per cent by 2023 (OECD/FAO 2014: 189).

In sum, irrespective of the growing purchasing power among the new middle classes and accounting for population growth, the national average per capita consumption rates in the Global South will remain consistently below the rates of most countries of the Global North. This is particularly true for beef, considered as the most harmful option in terms of its ecological footprint.

Meat consumption in individual countries

While international comparisons based on national averages on a per capita basis can shed light on the broad variety of dietary traditions and consumer preferences between different countries, they even out significant regional and social differences, each of them analysed in a vertical perspective by comparing different income levels, and in a horizontal perspective as reflecting different professional, educational and cultural patterns, and lifestyles co-existing within the framework of particular income levels. This is also true in regard to dietary traditions, as food consumption practices evolve in specific socio-cultural histories. The focus of this section is thus on individual countries, which will be discussed in three stages: 1) regional geographies of meat cultures; 2) the rural-urban setting of change; 3) cities as hotspots of changing consumer cultures. India and China are presented as the main case studies, as these countries most often come up in political publications that refer to soaring consumerism in the Global South, not least in relation to meat.

Regional geographies of meat cultures

There is not a single country without differing regional traditions, representing specific outcomes of an historical interplay of both normative (such as rules and institutions) and material factors (such as e.g. topology and infrastructure) (see Schatzki 2010) thus leading to particular cultural geographies. This applies not least in relation to how meat consumption is appreciated. In the following section this will be discussed on the example of how such trends evolved in both India and China.

The traditional geography of meat consumption in India[3]

Table 1.2 shows the average values for the different types of meat and aquatic products in India. The very low meat consumption as well as the composition of the Indian meat basket sum up the strong vegetarian tradition of Hinduism. According to official statistics, 80.5 per cent of the Indian population is Hindu, 13.4 per cent is Muslim, slightly more than 2.3 per cent are Christian and other religions make up an additional 3.7 per cent (Sikhs, Buddhists, Jains, other religions) (SECC 2011). Historical value systems, which are rooted in religion,

Table 1.2 Meat consumption in India, 2013, types of meat (Source: OECD Data 2015)

Meat consumption in India, 2013 (per capita kg/year)			
Beef	Pork	Poultry	Sheep
0.6	0.2	1.7	0.6

thus lead to stable rates of meat consumption, despite the profound changes in Indian society over the last century. Further historical factors are also important, sometimes to an even greater extent than religious traditions. This can be seen by distinct regional profiles of meat consumption, which prevail until today across India.

In some regions, the impact of non-Hindu traditions is obvious. In the case of Jammu and Kashmir, for example, the strong Muslim presence of more than 60 per cent is undoubtedly one reason for above-average values for the consumption of sheep and goat[4]; in Nagaland and other north-eastern territories the importance of pork points to the strong presence of Christian traditions (more than 80 per cent) and tribal religions. The lowest proportions of vegetarian families can be found in Kerala (2 per cent), West Bengal (3 per cent), Andhra Pradesh (4 per cent) and Orissa and Tamil Nadu (with 8 per cent each). But only in Kerala is the proportion of Hindus significantly below the national average (Hindu 52 per cent, Muslim 27 per cent, Christian 20 per cent). In the other four states, their proportion ranges from 73 per cent in West Bengal to 94 per cent in Orissa. The fact that these are all coastal states can explain the low proportion of vegetarians, as fish is easier to access than in land-locked states (Yadav/Kumar 2006) such as Rajasthan (63 per cent), Haryana (62 per cent), and Punjab (48 per cent) with their high proportions of vegetarian families. In fact, comparatively high consumption rates for fish exist in the coastal states: the highest in Kerala (2.3 kg per person/month), followed by West Bengal and the other three afore-mentioned coastal states, while the figures in land-locked states such as Rajasthan, Haryana, Punjab and others are in the range of less than 50g per person/month. The national average is around 260g (NSSO 2014).

The comparatively higher 'material' availability of fish near the coast may partially explain this disparity. However, it does not explain why, in most states of this group, the consumption of poultry with quantities of between 300 and 450g per person/month significantly exceeds the all-India average of around 200g, while remaining clearly below the national average in the states of the second group as well. The same applies to the consumption of eggs, where the second group of federal states also has much lower values. The differences in the regional distribution of religious affiliations are much less clear, in contrast.

Religious norms, also in relation to vegetarianism, have always demonstrated a certain elasticity[5]: the State of the Nation Survey 2006 showed that only 55 per cent of Brahmins, as the top caste, saw themselves as vegetarians in the narrower sense, in contrast to a still lower gauge of 12 per cent among the 'low castes and "outcastes"' (Yadav and Kumar 2006). The caste system in India is therefore also an indicator of what food is consumed. I will return to this point later in the chapter, in comparing rural and urban consumption according to income brackets. On the other hand, 8 per cent of the Christian population in India is vegetarian, although corresponding religious taboos play no role in this context. The Christian religion and tradition also knows meat-free days of the week and fasting periods, however. Yet a large number of people are satisfied with small amounts of meat even without those traditions, or without vegetarianism being an explicit point of reference for them at all.

In other words, vegetarianism is actually 'a function of inherited cultural practice rather than individual belief' (Yadav and Kumar 2006). Although current customs and traditions are characterised by religious traditions, in reality it is not solely the explicit rules that count, as is the case everywhere else in the world. Instead, it is the overall material, social and cultural framework in which the individuals were and are embedded in their daily life. Within this framework, decisions are taken as to which interpretations of traditional religious and other rules are accepted, interpreted and followed. As the case of India demonstrates, how food is practised can differ widely with regional patterns of meat consumption understood as a result of the interplay between these processes in the past. Individual preferences and consumer practices co-evolve with such structures.

The traditional geography of meat consumption in China

In China today, traditions of vegetarianism rooted in religion play a very limited role. Compared to India, China evolved historically to be at the opposite end of the spectrum: as shown below (Table 1.3), meat consumption in China – 49kg p.c./year compared to 3.1kg in India – is about sixteen times higher than in India.

Nevertheless, China is hardly any different from India in that very differing regional traditions in meat consumption are evident. Here as well, they have given rise to their own geography in relation to the particularly important types of meat in each case. The relatively low importance of religious norms emphasises the material roots of this geography all the more.

Despite wide overlaps, three main regions are evident, each of which has its own priorities in relation to meat products: there is a main area of beef and mutton consumption in the north and north-west: Tibet, Xinjiang, Ningxia and Inner Mongolia. In the south, both types of meat are represented much less. There, in contrast, the consumption of poultry is more important, particularly in Hainan, Guandong and Guangxi. China's most preferred meat

Table 1.3 Meat consumption in China, 2013, types of meat (Source: OECD Data 2015)

Meat consumption in China, 2013 (per capita kg/year)			
Beef	Pork	Poultry	Sheep
3.5	31.2	11.4	2.8

is pork, with a concentration of pork consumption in the country's south-west and south, especially Yunnan, Sichuan, Guizhou and again Guangdong. In the Xinjiang province, where Muslims form the largest population group, the consumption of pork meat is the lowest in the whole of China, however. Not surprisingly and in analogy to India, aquatic products (fish, shrimps and prawns, other aquatic products) are consumed more in China's coastal provinces, especially in the southern and south-eastern provinces, while aquatic products are least represented in the western and north-western provinces (with the greatest distances from the coasts) in particular (see Zhou et al. 2012 for a detailed overview).

The different spatial and topological demands for livestock farming are deemed to be the most important causes of the varying regional importance of beef and mutton on the one hand, and pork on the other: while traditional cattle and sheep farming requires extensive pastures, pig farming can be undertaken as backyard-centred household production because 'pigs are unfit for a pastoral way of life' (Hartog 2006: 21). However, this form of production has not been able to satisfy the current demand for some time, however. Suitable pastures for cattle and sheep farming were available only in certain regions, while others allowed for more intensive agricultural use of smaller areas. This type of farming requires cattle primarily as draught animals, which makes it counterproductive to slaughter them. In the past, the authorities even imposed penalties for killing cattle, which resulted in beef consumption becoming taboo in certain parts of China: but now not for religious, but rather material and pragmatic reasons. The effects of this cattle-protecting ideology have remained until today and have contributed to the formation of 'different food cultures' on the regional scale (Liu and Beblitz 2007: 4f).

These cultures, which have evolved historically and have roots in specific regions, develop a certain life of their own. They thus still retain their shaping influences even when the normative and/or material reasons for their historical emergence are losing their significance. Fitting into this framework is the fact that, in China, pork meat remains by far the highest in demand, while beef and mutton follow but with a significant gap.

The rural–urban setting of change

In addition to the geography of meat consumption, the rural–urban gap has to be taken into account as a second type of regional differentiation. While the former is primarily the result of different material and cultural circumstances in the past, the latter forms the framework within which new structures develop which will characterise future developments. The disparity between city and countryside has now lost much of its earlier unequivocal character: former exclusively urban functions have now spread into rural areas as well, particularly in the sector of non-agrarian production, and also in the sector of commerce and services. However, agricultural production remains one function that still structures rural areas to a great extent. This applies especially in the countries of the Global South. The cities, in contrast, now even more than in the past, are centres of a multitude of functions, including production, trade, social and medical services, education and training, media, culture and not least administration and political decisions. In this sense, cities are the most important centres of functional diversity, economic dynamics, social density and the exchange and meeting of national and international players and interests. They are thus breeding grounds of change. What is the impact on historically evolved traditions and cultures of meat consumption? Is historical diversity being replaced by greater homogeneity and are there new patterns of emerging diversity? As purchasing power is a significant variable, different income brackets are considered. In addition to red and white meat, the consumption of aquatic products continues to be taken into account.

Changes in the quantities consumed and preferences in India

For reasons of clarity, only six out of the full spectrum of twelve income brackets are considered here. The national average for 'fish and prawns' is 83g in rural areas and 252g in urban areas (Figures 1.4 and 1.5). No corresponding totals are listed for meat.

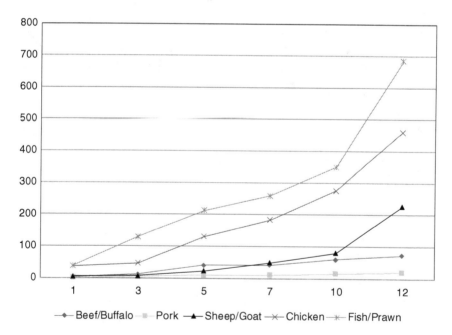

Figure 1.4 Consuming meat and fish in *rural* India; per capita gram/year.
Source: NSSO 2014, own diagram.

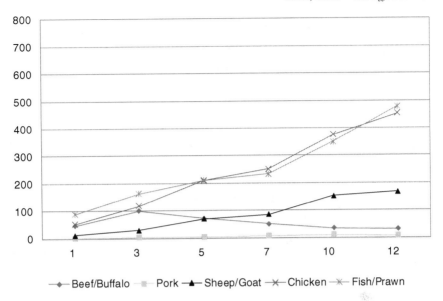

Figure 1.5 Consuming meat and fish in *urban* India; per capita gram/year.
Source NSSO 2014, own diagram.

The figures confirm two things. First, the consumption of *beef and pork* in rural areas as well as in urban areas is at the lowest level in all income brackets. The consumption of pork is always close to zero, in the countryside as well as in the city, for the poor as well as the prosperous. The consumption of beef and buffalo is below 100g per month everywhere. In rural areas, the lower the household income, the lower the consumption; in urban areas, the highest consumption value is to be found (albeit on a very low level as well) not for the income brackets 10–12, which have the highest income, but for income bracket 3, among poorer populations. This relates to beef and buffalo meat being particularly cheap in India. After his inauguration in 2014, Prime Minister Modi of the (Hinduist) Bharatiya Janata Party launched a political campaign to impose an even stricter ban on beef than already existed. A reaction, not least from prominent Hindus, was political resistance under the slogan 'I will eat beef, who are you to question me?' (Chief Minister of the federal state of Karnataka, *Hindustan Times*, 30 October 2015). *Sheep and goat* meat follow with a clear gap: in the urban sector with a continuous increase in higher income brackets, but in very low quantities; in the rural sector, a minimum quantity which can be distinguished from zero is only reached in the households which are strongest financially. *Chicken*, in contrast, is by far the number one. This applies for the whole spectrum of income groups. Chicken consumption is still exceeded by the consumption of aquatic products, however: in the urban sample only slightly; in the rural sample, in contrast, with a clear margin. Nevertheless, in 2011 in the rural sample, even in the top income bracket – all types of meat and aquatic products included – the total amounted on average to not even 1.5kg (1466g) per month and person.

Second, the consumption of meat increases with increasing income in all income brackets – with the exception of pork and beef, due to the associated taboos. This concerns in particular the upper half of the income brackets, albeit here as well with the qualification that the amounts consumed remain quantitatively still far below the corresponding quantities in other developing countries, let alone in most of the OECD countries. In relation to the lower half of the income brackets, the starting values are so low in absolute terms that even small increases are significant in percentage terms, especially in rural areas – another expression of the still prevalent poverty in India.

In summary, the figures do not provide any indication that with increasing household budgets a swing to more 'Western' meat preferences can be expected in the foreseeable future, both in terms of quantities and types of meat – particularly beef. At the same time, the clear preference for chicken is not solely a function of the relatively low price of this type of meat, as chicken is also the most popular category of meat in the most prosperous households – as is also the case in most countries of the Global North. Finally, no less remarkable is the fact that the consumption of fish in urban households is equal to that of chicken, and even exceeds it in rural households.

Changes in the quantities consumed and preferences in China

Compared to India, China also exhibits a number of similar trends despite the generally higher level of meat consumption.[6] In China, as well, higher incomes go hand-in-hand with an increase in meat consumption, in the countryside as well as in the city. And just like in India, the average incomes in the cities are at a much higher level than in the countryside. In China, after a significant increase compared to 2000, the average level of income in 2009 in the countryside amounts to 5153 yuan, where it is still below the urban average adding up to 6280 yuan in 2000. Accordingly, the growth rate of meat consumption in the countryside was much lower than in the cities: below 14 per cent (from 17.2kg to 19.6kg per person/year) in the countryside against 34 per cent in the urban sector (from 25.5kg to 34.67kg) (Zhou et al. 2012: 3).

In China as well, *poultry* achieved the largest increases in consumption in recent years. In rural areas, the consumption increased by 54 per cent: from 2.8kg to 4.3kg. In the same period, the consumption in the urban sector increased by 94 per cent, and this was from a much higher starting level: from 5.4kg in 2000 to 10.5kg in 2009. The same pattern is evident for *aquatic products*: an increase of slightly more than 34 per cent in the rural sector and 22 per cent in the urban sector, but in the urban sector to more than twice the quantity in the countryside – to 14.3kg compared to 5.3kg. Yet, the higher the incomes the more the share of food expenditure out of total living expenditure has nevertheless decreased: from 49 per cent to 41 per cent in the rural sector and from 39 per cent to 36 per cent in the urban sector. Comparing the lowest and the highest of five income brackets, the change is even clearer, and most pronounced in the urban sector (Zhou et al. 2012: 5, 9–10).

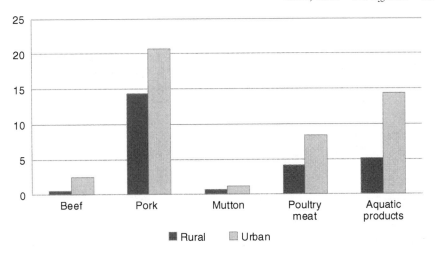

Figure 1.6 Consuming meat and aquatic products: rural and urban China, 2009; per capita kg/year.
Source: Zhou et al. 2012: 15; own diagram.

In the red meat sector, *pork* is still number one (Figure 1.6). In the countryside, the increase is minimal in the period under comparison, however. In the urban sector, pork yet again exhibits an increase of more than 20 per cent despite its much higher starting level. Although *beef and mutton* also show increases, both remain far behind poultry and aquatic products: beef at 0.56kg in the rural sector and 2.4kg in the urban sector, and mutton at 0.8kg in the rural sector and 1.3kg in the urban sector. Here as well is a parallel to India, together with the comparatively strong demand for poultry and aquatic products.

Last but not least, the strong differentiation of the quantities consumed according to income brackets is also repeated. It is the top four of the five different income brackets which, in the urban sector, demand more pork as well as more beef, in the case of beef the demand is no less than 50 per cent higher than in the lowest income bracket. In addition, comparing income brackets produces two patterns: while the differences between the income brackets in both countryside and city are slight for pork and mutton, they are very striking for poultry and aquatic products. In both product groups, the quantities consumed in the upper income bracket are over 100 per cent more than those of the lowest group. Moreover, in the countryside, the top income bracket is yet again markedly different from the spectrum of the other four groups. Here, poultry and aquatic products are clearly the top favourites of the most prosperous consumers, even more so than in the urban sector (Zhou et al. 2012, 18–20)

Cities as hotspots of changing consumer cultures

Urbanisation is now far advanced, and it will continue to advance in the future. Cities are distinguished not only by their size, but also by the diversity and volume of economic, social, cultural and politico-administrative services available (Kamal-Chaoui et al. 2009). In comparison to rural areas, most cities are more complex in infrastructural, economic, social, cultural and political respects. These characteristics are, on the one hand, a precondition for a broad spectrum of employment and related income opportunities (Dobbs et al. 2012). On the other hand, they provide a specific framework within which the inhabitants have to go about their daily lives.

In this sense, cities also offer a broader range of products and product variations in relation to diet. Cities have differentiated and efficient systems of provision in the wholesale and retail sectors. They allow each consumer to make a personal choice without having to spend a significant amount of time doing so. On the other hand, citizens must adapt their daily life to the special features of the urban structures: this applies to the widespread separation of place of residence and place of work and the associated demands placed on their mobility. If both partners in a household are in employment, this often goes hand-in-hand with long, strictly regulated and unequal working hours. This aggravates the problems of time management. As a result, the opportunity for sharing meals together decreases, and cramped living conditions reduce the possibilities to cook and eat, especially with friends.

Eating out and the use of pre-cooked meals and convenience food turn out to be the fitting answer to such specifically urban constraints and opportunities. While opportunities for eating out matured in more developed countries and were then exported to the cities of the Global South by Western companies, they nevertheless represent primarily urban answers to specific urban challenges and possibilities. In that sense, the rapid spread of such forms of eating to all large cities in the world is less related to the capitulation to Western patterns of taste or the business model of 'McDonaldization' described by Ritzer (2011), rather than to its suitability for urban customers under the conditions of today's urban reality.

It does not appear easy to correctly assess the extent of meat consumption away from home, however. The FAO does not display data on this issue. National statistics provide indications, but these figures are not very differentiated (several chapters in this volume point to trends towards eating out across the region). As for *India*, the NSSO classify 'cooked meals purchased' according to frequency during the past 30 days, in 'rural' and 'urban' contexts and according to income brackets (Figure 1.7). The survey shows very low average frequencies for the survey period: 0.42 (rural) and 1.54 (urban) cases respectively per 30 days.

According to the survey, eating away from home is primarily experienced by those in the top income brackets. Since these are average figures, however, a considerable proportion of the people in these brackets are eating out much more frequently. Since the Indian NSSO measures the meals away from home in frequencies, meals taken from small cook shops and street kitchens should have been taken into account. This would make the low frequencies in the bottom half of the income brackets all the more remarkable. The figures show simultaneously the extent to which eating away from home is first and foremost a characteristic of urban ways of life.

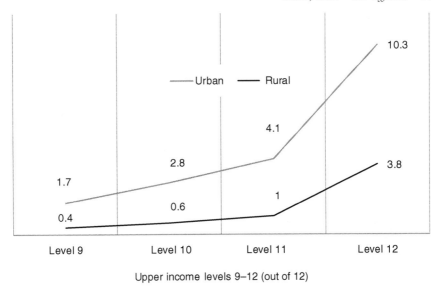

Figure 1.7 Frequencies of cooked meals purchased; per capita/month (India).
Source: NSSO 2014, own diagram.

Zhou et al. (2012) show a very similar picture for urban *China,* yet on a higher level. The average proportion of the food-away-from-home expenditures in the total food expenditures increased from 10 to 22 per cent between 1995 and 2009, which most likely also contributed to the increase in meat consumption. The vast difference between the high-income consumers and the lower-income groups, evident in India, is again repeated here. Expenditure on away-from-home food by the upper income group (out of ten groups) in 2009 exceeded the total food expenditure by the lowermost income group by 25 per cent. While the wealthy consume increasingly out of the home, the lower income groups are also spending more on food, as their income situation has also significantly improved on average, yet at a significantly slower pace. At the same time, the gap between the respective income brackets in rural areas and between rural and urban areas has increased. The evident problem of lacking equity and justice is one of the key challenges confronting the political decision-makers in China (cf. OECD 2013).

For many urban consumers, however, eating out is both a matter of preference and symbolic social distinction as well as a way to meet specific urban requirements and possibilities for managing their daily life. These new patterns of consumption represent a shift from both the cultural patterns of consumption of their earlier years, as well as rural life traditions. Does this imply that India, China and the large number of comparable urban regions of the Global South have set out down a path leading to Western cultural domination? There is no lack of fierce criticism in this regard, especially in relation to the younger generation. A journalist expressed this criticism in the context of India: 'They adopt anything and everything that has a Western label.' In a sweeping statement, the author contends that the even larger group of 'urban Indians' has a 'predilection … to eat junk food, deep-fried in transfat' (D'Monte 2011). In a criticism informed by modernisation theory, the Indian social scientist Gupta calls it in general a 'westoxicated elite' (Gupta 2007).

Illustrative examples can certainly be provided to support this criticism, yet how representative such examples are remains unclear. The fast food sector in India provides a particularly instructive example of how strongly the traditional culture of one's own country influences this consumer goods sector. Although McDonald's is omnipresent in India's larger cities, as in most other cities worldwide, the management quickly realised that habitual profits would not be garnered through beef-centred burgers, as is the case with the classical, 'Western' version. Hence, the menu today largely comprises no-beef options, from chicken and fish to eggs and completely vegetarian options. Since 2012, completely vegetarian McDonald's restaurants have also been established. Moreover, on the company's website, 'nutritional information' is provided for all meals, including the percentage of transfat and cholesterol, most likely more of a concern to 'modern' educated urban customers.

In China, although all types of meats are popular, traditions are also being challenged through specific forms of consumption and associated references. Zhou et al. (2012) point out that eating together with family and friends is a tradition in China, with the variety of dishes available a sign of prestige. Types of meats are not given hierarchical significance in terms of status, as is the case in other countries such as the social significance of beef in the USA and countries in South America. Moreover, 'trying foods from other cultures is popular' as well (Zhou et al. 2012: 42). The broad range of foods offered by the large supermarkets opens up new possibilities to satisfy this interest. Here as well, special interest is paid to products from other countries, Western countries included. The preference for Western products includes not only expensive branded goods, but also products consumed on a daily basis. Health is a crucial reason for this preference. After serious food scandals, primarily due to contaminated milk powder, food safety turned into a highly important topic in China – in dynamic interaction with widespread general distrust in Chinese economic, administrative and political officials, deemed irresponsible and corrupt (Yang 2015: 186). Western standards and associated food products are perceived as more reliable, in contrast.

Although higher incomes translate into increases in all forms of meat, as well as eggs, dairy products and fish, beef has remained quantitatively less important (Figure 1.8). Beef is not coveted by many Chinese people, who do not appreciate the smell and taste, or are unsure as to how it is to be prepared. The result is that many prefer to eat beef in a restaurant – as one dish among many. Most of the increases in beef consumption are expected from upmarket hotels, high-income Chinese consumers, expatriates and tourists (Zhou et al. 2012: 61).

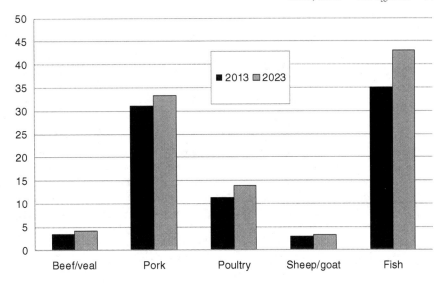

Figure 1.8 Consumption of meat and fish in China, 2013–2023; kg per capita/year.
Source: Zhou et al. 2012, own diagram.

The NMCs – drivers of change?

Which conclusions can be drawn about 1) the role of the NMCs as drivers of consumption and 2) trends of meat consumption in particular?

Considering that, per definition, the NMCs are those between the rich and the poor, two consequences can be derived in advance: the NMCs are not 'the rich' or 'the new rich' or 'the affluent', as suggested implicitly or explicitly by scholars (e.g. Robison 1997; de Zoysa 2011) and by authors of prominent books about present cultural trends in the South (e.g. for India: Dasgupta 2014). Second, the NMCs are far from representing a homogenous social actor (Birdsall 2015) that would justify addressing them in the singular as 'the new middle class' (see e.g. Varma 1998, 2014) or 'the consuming class' (Dobbs et al. 2012).[7] The distribution of meat consumption according to income brackets as shown above can be seen as a perfect illustration of the gradations between the bottom, middle and top sections of the middle classes.

The focus of much of the ecological and social concern put forward by social scientists is on the upper third tier of the ladder (e.g. Myers and Kent 2004, Fernandes and Heller 2009; Mawdsley 2009) while there are fewer studies that describe the daily reality of life in the lower sections of the NMCs, such as for example Nischalke's detailed study of food habits among the lower middle classes in the Indian mega city of Hyderabad (2013). This applies not least to the extremely small quantities of meat consumed here, which is fully in line with the data on very different levels of meat consumption in India and China as set out above.

However, even in the upper tiers, there is no homogeneity in terms of attitudes and/or behaviour. While parts provide ample evidence of substantial ignorance thus confirming the bad guy image – Varma (1998) coined the harsh term 'consumerist predators' – others express significant civic and environmental concern: from attitudes to activism and, depending on the specific topic at stake (e.g. Lange et al. 2009; see also Varma 2014), partial overlaps of concern between people belonging to fairly different income brackets and milieus (Leiserowitz et al. 2013).

This is very much in line with those patterns that research on environmental attitudes and behaviour as well as on unsustainable/sustainable consumption have set forth recurrently with a focus on consumers in the Global North (see e.g. Eurobarometer 2009; Warde 2005; Defila et al. 2012) – with the only difference being that even in the upper tiers of the Southern middle classes, consumption is and will continue to lag behind that of their Northern classmates. Nevertheless, also with regard to consumption, as in the Global North, particularly the upper tiers of NMCs in the Global South still are, and will become even more so, economic, social and political key actors anyway, some for the better and some for the worse. They are thus 'both part of the problem and of possible solutions for the world's developmental challenges' (Knorringa and Guarin 2015: 202).

As for meat in particular there are a number of characteristics which allow for changes in traditional habits on a step-by-step pace. Different types and quantities of meat can be tested easily, they can be combined with other foods, and in relation to the price of a meal. Hence, alternatives to traditional patterns of meat consumption can lead to a broad range of hybrid solutions, not least including elements of different national and international traditions from urban as well as rural areas. As Seubsman et al. expose, this applies not least to fast food (2009). There is also a broad range of possibilities for trying out new paths on the basis of smaller meat quantities or increased vegetarian options. Thus, new trends in lifestyles and fashions can become templates for fostering not only changes which are problematic in terms of overusing resources but of changes with contrary effects, too. The different spending powers and the various cultural orientations within the NMCs can serve as a favourable starting point.

All in all, given this variety of options and considering that, in developing countries including those with the most booming economies, the quantities of meat consumed per capita continue to remain below those of the rich countries, even in the higher income groups of the NMCs, the heated debate about increasing rates of meat consumption in the Global South appears exaggerated and even misleading. For the foreseeable future at least, there is little cause for sweeping moral reproaches and dramatic alarm calls because of alleged consumerist excess and lack of ecological responsibility in relation to the NMCs. For the majority of this group rising purchasing power means not affluence but first and foremost increasing choice, which is one essential prerequisite for development and social welfare (HDR 2013).

The big problem, however, derives from the fact that even small increases in the per capita consumption of meat will increase the quantities needed and their ecological footprint in a way that does, indeed, give rise to serious concern (FAO Statistical Yearbook 2013), given the large number of people who benefit from increasing purchasing power, though at mostly modest pace and levels. This challenge primarily requires a 'reasonably equitable national and international burden sharing' (Wiemann 2015: 210).

Millennium Consumption Goals (MCGs) in analogy to the Millennium Development Goals (MDGs) of the United Nations could be a model (cf. de Zoysa 2011). MDGs are set to open trajectories to consumption options, which involve less resource intensity. But this needs to be done in a way that meets the demands of two different addressees. In the Global South, in particular, low-income groups need to increase their economic and social participation, as well as gain from rises in purchasing power. And in both the Global North and the Global South, curbing the level of resource-consuming practices has to gain acceptance among the better-off sections of society, albeit on different scales (see Di Giulio and Fuchs 2014; Spangenberg 2014). This is a tough agenda. 'Coherent policy frameworks' (Reisch et al. 2013; see also OECD 2010) are a basic prerequisite for getting ahead, and the policies needed will hardly take shape without unequivocal political support in political-administrative respect, on the one hand, and substantial involvement of the civil society on the other. Professional expertise and support from significant sections of the NMCs have to be a cornerstone. At the same time, politics has to prevent them from thwarting the needs and demands of the poor and of the less favoured sections of the middle classes (OECD 2012), resulting in sustainable consumption solely for an elite. This applies to both the Global North and South, yet with substantial differences (Schmidt 2009) in terms of temporal, economic, and cultural frameworks and needs – as the saying goes in some parts of Asia, *'same, same but different'*.

Notes

1. As referring to the emergence of and new trends in the composition of modern societies, the term NMCs is a constituent part of the sociological tradition (see Lange and Meyer 2009, 6–12; López and Weinstein 2012). Its meaning in the present debate on consumption, however, is predominantly shaped by market research, revolving around one core question: how does the increasing purchasing power of households spread across different income brackets, located above the poor and below the rich (e.g. Kharas 2010, 10)? The term 'new' middle classes emphasises that the purchasing power under consideration has originated only in the course of the last ten to fifteen years.
2. 1Mt equals 1,000,000 metric tons.
3. The basis of the data presented here is the Indian *National Statistical Survey* (NSSO 2014). It presents representative and detailed data on the development of nutritional intake and household consumer expenditure across socio-economic groups, both in rural and urban India. The survey was stratified into twelve income brackets. Separate samples were taken in all Indian federal states. Further data is provided by the *Socio-Economic and Caste Census 2011* (SECC 2011) and the *State of the Nation*

Survey 2006 (Hindu-CNN-IBN 2006). See also *The Anthropological Atlas* (Singh 1993).
4   Unlike the nomenclature used by the FAO, the NSSO refers to 'sheep and goat', 'beef and buffalo', and 'fish and prawns'.
5   In their survey, the authors of the *Anthropological Atlas* found many other groups in addition to pure vegetarians, including those who eat fish, others who eat fertilised or non-fertilised eggs. Some avoid *masur dal* or lentil, some do not eat onions, while others avoid garlic (Singh 1993).
6   In addition to the data provided by FAO and OECD, in the following, the publication of Zhou and colleagues on food consumption trends in China is given priority. The authors build on both FAO and OECD data and on data provided by China's National Statistical Bureau (Zhou et al. 2012: 72). It must be noted, however, that these figures show significant differences compared to those provided by FAO and OECD. The authors touch the issue explicitly (pp. 8, 47): their main concern is that the data available from China's State Statistical Bureau 'include only the quantities of foods purchased by consumers for consumption at home', which might well entail a bias, considering the rapidly growing importance of *eating away from home*, particularly in larger cities. This can help to explain the diverging figures on meat consumption provided by FAO and OECD data and China's Statistical Office. If there are further aspects to be considered they cannot be discussed here. At the same time, the Chinese data is very helpful. Different from FAO/OECD data in the same way as data provided by the Indian NSSO, the Chinese data provides time series on meat consumption on a household basis according to income brackets and for both rural and urban areas. As a distinct statistical framework, they permit a more detailed understanding of some of the dynamics of meat consumption.
7   Kharas, for example, includes households (four persons) with daily expenditures between USD 10 and USD 100 per person in purchasing power parity (Kharas 2010, 12 f. including an appraisal of the limits of this kind of yardstick); see also Myers and Kent (2004).

References

Birdsall, N. 2015. 'Does the rise of the middle class lock in good government in the developing world?' *European Journal of Development Research* 27: 217–229.
Census 2001: http://censusindia.gov.in/Census_Data_2001/Census_data_finder/C_Series/Population_by_religious_communities.htm (accessed 20 August 2015).
Dasgupta, R. 2014. *Capital: The Eruption of Delhi*. New York: Penguin Press.
Dauvergne, P. 2008. *The Shadows of Consumption: Consequences for the Global Environment*. Cambridge, MA: MIT Press.
Defila, R., A. Di Giulio, and R. Kaufmann-Hayoz (eds). 2012. *The Nature of Sustainable Consumption and How to Achieve It*. Munich: Oekom.
D'Monte, D. 2011. 'One man's meat is simply another's poison.' Asian Conversations. January; http://www.asianconvewrsations.com/IndiaNonVeg.php (accessed 26 June 2015).
Dobbs, R., J. Remes, J. Manyika, C. Roxburgh, S. Smit, and F. Schaer. 2012. *Urban World: Cities and the Rise of the Consuming Class*. The McKinsey Global Institute: McKinsey. www.mc mckinsey.com/mgi/
Eurobarometer. 2009. 'Europeans' attitudes towards the issue of sustainable consumption and production.' EB Series #256, Conducted by the Gallup Organisation, Hungary at the request of the European Commission. Directorate-General for the Environment, Brussels.

FAO Statistical Yearbook. 2013. www.fao.org/docrep/018/i3107e/i3107e00.htm. (accessed 15 October 2015).
Fernandes, L. and P. Heller. 2009. 'Hegemonic aspirations: New middle class politics and India's democracy in comparative perspective.' In *Whatever Happened to Class? Reflections from South Asia*, edited by R. Agarwala and R. J. Herring, 146–165. Lanham, MD: Lexington Books.
Giulio, A. Di, and D. Fuchs. 2014. #Sustainable consumption corridors: Concept, objections, and responses.' *GAIA* 23 (S1): 184–192.
Gupta, D. 2007. *Mistaken Modernity: India between Worlds*. New Delhi: HarperCollins India.
Hallström, E. and P. Börjesson. 2013. 'Meat-consumption statistics: Reliability and discrepancy.' *Sustainability: Science, Practice, and Policy* 9(2): 37–47.
Hartog, A. den, W. van Staveren, and I. Brouwer. 2006. *Food Habits and Consumption in Developing Countries: Manual for Social Surveys*. Wageningen: Wageningen Academic Publishers.
HDR (Human Development Report). 2013. *The Rise of the South*. New York, NY: United Nations Development Programme.
Hindu-CNN-IBN 2006: State of the Nation Survey 2006. www.mcdonaldsindia.net/burgers-and-wraps.aspx (accessed 7 June 2016).
Hindustan Times, Bengaluru, 30 October 2015. www.hindustantimes.com/india/i-will-eat-beef-who-are-you-to-question-me-karnataka-cm/story-dWpJhRO8GlEaEd4oaCP8mJ.html (30 October 2015).
Kamal-Chaoui, L., E. Leman, and Z. Rufei. 2009. 'Urban trends and policy in China.' *Working Papers 2009 (1)*. Paris: OECD Publishing.
Kharas, H. 2010. 'The emerging middle class in developing countries.' *Working Papers*. Paris: OECD Development Centre.
Knorringa, P., and A. Guarin. 2015. 'Inequality, sustainability and middle classes in a polycentric world.' *European Journal of Development Research* 27(2): 202–204.
Lange, H. and L. Meier. 2009. 'Who are the new middle classes and why are they given so much public attention?" In *The New Middle Classes – Globalizing Lifestyles, Consumerism, and Environmental Concern*, edited by H. Lange and L. Meier, 1–28. Dordrecht: Springer.
Lange, H., L. Meier, and N. Anuradha. 2009. 'Highly qualified employees in Bangalore/India: Consumerist predators?' In *The New Middle Classes – Globalizing Lifestyles, Consumerism, and Environmental Concern*, edited by Hellmuth Lange and Lars Meier, 281–298. Dordrecht: Springer.
Leiserowitz, A., J. Thaker, G. Feinberg and D. Cooper. 2013. *Global Warming's Six Indias. Yale Project on Climate Change Communication*. New Haven, CT: Yale University.
Liu, H. and C. Beblitz. 2007. *Determinants of Beef Consumption in China*. Charles Sturt University (Australia): Asian Agribusiness Research Centre, Working Paper 41.
López, R. and B. Weinstein (eds). 2012. *The Making of the Middle Class*. Durham and London: Duke University Press.
Mawdsley, E. 2009. '"Environmentality" in the neoliberal city: Attitudes, governance and social justice.' In *The New Middle Class. Globalizing Lifestyles, Consumerism and Environmental Concern and the Globalizing City*, edited by H. Lange and L. Meier, 237–252. Dordrecht: Springer.
Myers, N. and J. Kent. 2004. *The New Consumers: The Influence of Affluence on the Environment*. Washington, DC: Island Press.

Nischalke, S. 2013. *Changes and Risks in the Mega-urban Food System of Hyderabad/India.* Göttingen: Goltze.

NSSO (National Statistical Survey). 2014. *Household Consumption Survey 2011–12.* 68th Round, edited by National Sample Survey Office. New Delhi: Government of India, Ministry of Statistics and Programme Implementation.

OECD. 2010. *Consumer Policy Toolkit.* Paris: OECD Publishing.

OECD. 2012. *Poverty Reduction and Pro-Poor Growth: The Role of Empowerment.* Paris: OECD Publishing.

OECD. 2013. *OECD Economic Surveys: China 2013.* Paris: OECD Publishing.

OECD Data 2015. *Agricultural Output.* Paris: OECD Publishing. https://data.oecd.org/agroutput/meat-consumption.htm (accessed 15 October 2015).

OECD/FAO. 2014. *Agricultural Outlook 2014–2023.* Paris: OECD Publishing.

OECD/FAO. 2015 *Agricultural Outlook 2015.* Paris: OECD Publishing.

Reisch, L., U. Eberle and S. Lorek. 2013. 'Sustainable food consumption: An overview of contemporary issues and policies.' *Sustainability. Science, Practice, and Policy* 9(2): 7–25.

Ritzer, G. (ed.). 2011. *The McDonaldization of Society,* 6th edition. Thousands Oaks, CA: Pine Forge Press.

Robison, R. (ed.) 1997. *The New Rich in Asia: Mobile phones, McDonald's and Middle Class Revolution.* London and New York: Routledge.

Said, E. W. 1978. *Orientalism.* New York: Pantheon Books.

Schatzki, T. 2010. 'Materiality and social life.' *Nature and Culture* 5(2): 123–149.

Schmidt, V. 2009. 'Convergence and divergence in societal modernization: Global trends, regional variations, and some implications for sustainability.' In *The New Middle Classes: Globalizing Lifestyles, Consumerism and Environmental Concern,* edited by H. Lange and L. Meier, pp. 29–48. Dordrecht and London: Springer.

SECC. 2011. *Socio-Economic and Caste Census 2011.* http://secc.gov.in/reportlistContent (accessed 20 April 2016).

Sedlacko, M., L. Reisch, and G. Scholl. 2013. 'Sustainable food consumption: When evidence-based policy making meets policy-minded research – Introduction to the special issue.' *Sustainability, Science, Practice, and Policy* 9(2): 1–6.

Seubsman, S., M. Kelly, P. Yuthapornpinit and A. Sleigh. 2009. 'Cultural resistance to fast-food consumption? A study of youth in North Eastern Thailand.' *International Journal of Consumer Studies* 33: 669–675.

Singh, K. 1993. *An Anthropological Atlas: Anthropological Survey of India.* People of India National Series, Vol. XI. Delhi: Oxford University Press.

Spangenberg, J. 2014. 'Institutional change for strong sustainable consumption: Sustainable consumption and the degrowth economy.' *Sustainability: Science, Practice, and Policy* 2(10): 62–77.

SSB (State Statistical Bureau). *Statistical Yearbook,* various issues. Beijing: China Statistical Press.

Tukker, A. 2008. *Perspectives on Radical Changes to Sustainable Consumption and Production.* Sheffield: Greenleaf.

Varma, P. 1998. *The Great Indian Middle Class.* London: Penguin Books India.

Varma, P. 2014. *The New Indian Middle Class. The Challenge of 2014 and Beyond.* Noida: HarperCollins Publishers India Press.

Warde, A. 2005. 'Consumption and theories of practice.' *Journal of Consumer Culture* 5(2): 131–153.

Wiemann, J. 2015. 'The new middle classes: Advocates for good governance, inclusive growth and sustainable development?' *European Journal of Development Research* 27(2): 195–201.

Yadav, Y. and S. Kumar. 2006. The food habits of a nation. *The Hindu*, 14 August. www.thehindu.com/todays-paper/the-food-habits-of-a-nation/article3089973.ec (accessed 1 July 2015).

Yang, X. 2015. *Megatrend: Nachhaltiger Konsum. Eine explorative Studie des chinesischen Biokonsums*. Diss., Göttingen.

Zhou, Z., W. Tian, J. Wang, H. Liu and L. Cao. 2012. 'Food Consumption Trends in China. April 2012.' Report submitted to the Australian Government Department of Agriculture, Fisheries and Forestry. www.agriculture.gov.au/SiteCollectionDocuments/agriculture-food/food/publications/food-consumption-trends-in-china/food-consumption-trends-in-china-v2.pdf (accessed 15 October 2015).

Zoysa, U de. 2011. 'Millennium Consumption Goals: A Fair Proposal from the Poor to the Rich.' *Sustainability: Science, Practice, and Policy* 7(1): 1–5.

2 Grappling with impacts

Environmental considerations around agriculture and food waste treatment in Asian cities

Megha Shenoy

Emerging countries in South and Southeast Asia – the epicentre of the 'new consumer'[1] phenomenon – are experiencing an increase in consumption like never before (Myers and Kent 2004: 7). This rise in consumption, which includes food, has undeniable adverse environmental and social impacts. A deeper understanding of local practices and policies[2] across the entire life cycle of food can positively influence sustainability[3] in the region. Currently, policies in most Asian countries are usually framed based on those drafted in developed countries, and are carried out by government ministries that are interconnected in the real world through linkages in resource consumption and disposal, but disjointed in their administrative duties. The life cycle of food illustrates these interconnections: for example, organic fertilizer (such as compost) that comes from food waste can be used in organic agriculture. But waste management and agriculture are usually governed by separate ministries that do not collaborate on building and enforcing policies that interconnect these areas. A life cycle approach that takes into account local practices and encourages cohesive policies is the need of the hour.

The life cycle of food includes stages from agriculture, to storage and distribution (wholesale and retail), to consumption, food wastage and its treatment. In this chapter, I focus on two life stages – agriculture and food waste treatment – that are of significance among urban residents, policymakers and practitioners in various Asian cities. Globally, there is a rise of organic produce availability and its perceived health benefits (Hughner et al. 2007: 8). Food waste treatment is a recent challenge due to the sudden expansion of cities and the dearth of space for landfills. Identifying low-impact options at both these life stages can significantly reduce emissions to air, water and soil.

First, a literature review of Life Cycle Assessment (LCA) studies that examine the potential environmental impacts of alternative strategies for agriculture (organic vs. non-organic), and for treating food waste (anaerobic digestion, compost production, incineration and landfill) is provided. Second, from this review general trends about environmental impacts that can inform local practices and policies on how to improve sustainability of food production and consumption are summarized. Third, practices and policies around the life stages of agriculture and food waste treatment are examined in the context of four Asian cities: Bengaluru, Kuala Lumpur, Metro Manila and Seoul. This step allows for placing the LCA literature review within specific contexts to recommend tailored low-impact options and to cross-educate policies.

An LCA is one of the most comprehensive methods to: i) identify the most significant potential environmental impacts (including Green House Gas [GHG] emissions, soil toxins and water toxins) along the entire life cycle of a product (from cradle to grave), and ii) evaluate trade-offs in potential impacts of comparable, alternative strategies at any one life stage (for example organic vs. non-organic agriculture) (Guinée 2002: 407). LCAs are extremely data intensive and require detailed quantitative production inventories (called life cycle inventories [LCIs]) of all inputs and outputs for each process along the entire life cycle of a product (Guinée 2002: 9). These LCIs, which are detailed matrices of flows from and to the environment (of water, energy, materials; see Figure 2.1), are specific to each country and production method. For example, the LCIs for corn cultivation in India and Europe will be very different. In India cattle are used for ploughing soil and manual labour is needed for harvesting crops, whereas in Europe tractors and combine harvesters that run on fossil fuels are used. Further, electricity that is used to power water pumps and other agricultural processes depends on different mixes of fuels (coal, hydro, wind etc.) in each country. Subsequently, the impacts of corn cultivation in these two countries will differ significantly. In addition, LCA involves a number of technical assumptions and implicit value choices (Guinée 2002: 426). Therefore, the exact results of LCAs are context specific and cannot be directly extrapolated from one location to another, even for the same product.

Completing an LCA allows one to assess the potential environmental impacts that a product may have on climate, human health, ecosystem quality and resource constraints. Undertaking a full LCA for various food products at the life stages of agriculture and food waste treatment in the context of most developing cities is premature as of 2015, because very little LCI data exists in these countries.

The recent popularity of LCA in the policy context (Wardenaar et al. 2012: 1059) has led to initiatives in developing countries to compile LCI datasets. Ecoinvent (an LCA database provider) has collated 150 LCIs for India, 82 for Malaysia, 8 for the Philippines, and 179 for South Korea (Ecoinvent 2015). These LCIs are still insufficient to carry out robust LCAs of agriculture and food waste treatment systems in these countries, as most of these datasets are adapted from global LCIs and not from primary data (Safaei 2015). In order to overcome the problems of data insufficiency and limitations of extrapolating the results of an LCA from one region to another, I have compiled a review of several LCAs so as to uncover some common patterns and general trends.

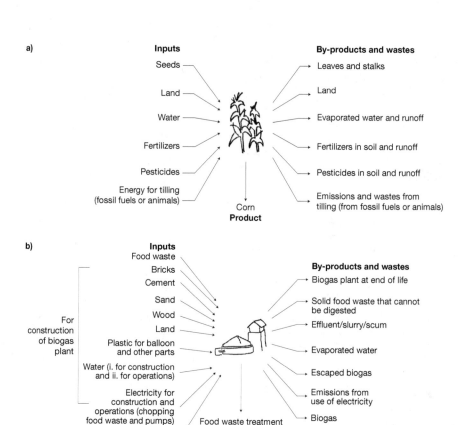
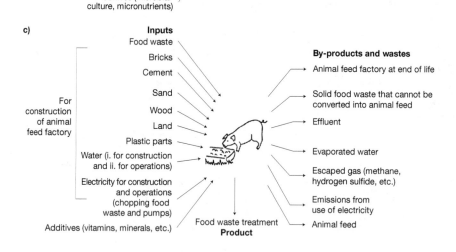

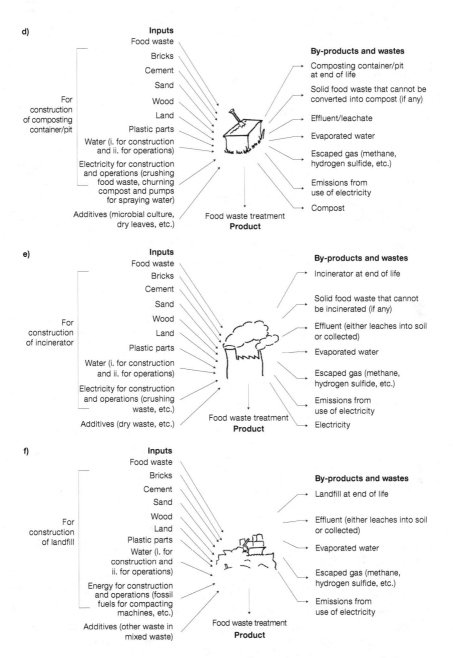

Figure 2.1 LCA inputs and outputs for agriculture and food waste treatment. Illustrations for a) corn agriculture, b) biogas from food waste, c) animal feed from food waste, d) compost from food waste, e) incineration of food waste, f) landfilling of food waste

Drawings by Megha Shenoy.

A comprehensive review of over 150 LCA studies of food products, from all over the world (largely from industrialized countries as most LCA studies are from these countries), has identified certain trends to lower environmental impacts:

1. In general, avoid air-transported products, prefer organic products and reduce meat consumption (Jungbluth et al. 2000, Gomiero et al. 2011: 137, 138).
2. Grazing animals (cows, sheep, goats) have higher environmental impacts compared to poultry and pork, as the overall impact of meat products is primarily governed by agricultural production of feedstock and the efficiency of conversion of this feedstock to animal mass (Jungbluth 2000: 143).
3. The total environmental impact of vegetables may vary by a factor of nine whereas that of meat may vary by a factor of seven, depending on geography, climate, soil characteristics, crop rotation cycles and inputs for cultivation (Jungbluth 2000: 143).

What follows is an overview of general impacts on agriculture, along with trends across Asia, and context-specific examples on agriculture practices and policies in Bengaluru, Kuala Lumpur, Metro Manila and Seoul. A similar approach is then taken to exploring food waste treatment impacts.

Impacts of agriculture: organic vs. non-organic

Agriculture contributes significantly to climate change (26–35 per cent of all anthropogenic GHG emissions) primarily due to greenhouse gasses that are emitted during crop production and livestock rearing (IPCC 2007: 3). Climate change in turn will increase the frequency and intensity of environmental threats making it harder to grow food in the future, adversely impacting human health (Sachs 2014). This codependence illustrates the vicious cycle between climate change and human actions (such as agriculture) that cause it. Environmental impacts of agriculture are primarily due to methods of the green revolution which began in the 1960s. This green revolution introduced farming methods that are characterized by the extensive use of chemical pesticides, fertilizers, external energy inputs, uniform high-yield hybrid crop seeds; single crops/row crops grown continuously over many seasons; large capital investment in production and management technology; high labour efficiency and, in the case of livestock production, meat that comes from confined and concentrated systems (Gold 1999). The green revolution accounted largely for increases in food production in developing countries; from 1961 to 1980 the growth rate of the yield of crops in Asia increased by more than three times (Evenson and Gollin 2003: 760). Agriculture popularized by this green revolution is also referred to as non-organic or conventional agriculture.

On the other hand, organic agriculture is defined by the International Federation of Organic Agriculture Movements (IFOAM) as:

a production system that sustains the health of soils, ecosystems and people. It relies on ecological processes, biodiversity and cycles adapted to local conditions, rather than the use of inputs with adverse effects. Organic agriculture combines tradition, innovation and science to benefit the shared environment and promote fair relationships and a good quality of life for all involved.

In general, evidence is accumulating about the significant reduction in overall environmental impacts of organic agriculture (Gomiero et al. 2011: 96–117; Ramesh et al. 2005: 561–567). A review of over 300 published reports (Stolze et al. 2000) showed that of the eighteen environmental impact indicators, organic farming systems performed significantly better in twelve and worse in none (Ramesh et al. 2005: 564). The primary reasons for the lower environmental impacts of organic farming (Lockeretz et al. 1981: 546; Pimentel et al. 2005: 575–581; Refsgaard et al. 1998: 611–627; Gomiero et al. 2011: 100–117) are due to:

1 lack of synthetic nitrogen fertilizers that consume ammonia which is manufactured from fossil-fuel-based methane;
2 lack of synthetic pesticides and herbicides;
3 low inputs of other mineral fertilizers (to supply phosphorous [P] and potassium [K]);
4 lower use of foodstuffs (concentrates) that are highly energy-consumptive.

Evidence is also accumulating about the long-term decline in productivity of non-organic agriculture, due to the adverse effects of agrochemicals on the health of soil, water, local pollinator and other beneficial fauna populations (Rao 1999). A global dataset of 293 cases shows that the average yield of food grown organically was higher than that grown non-organically in the developing world (Badgley et al. 2007: 86). Another survey of 208 projects in developing tropical countries found an average yield increase for organic crops of 5 to 10 per cent in irrigated land and of 50 to 100 per cent in rain-fed land (Pretty and Hine 2001: 13). Organic agriculture is therefore highly relevant in arid, semi-arid and water-stressed regions because non-organic practices can erode and contaminate fertile top soil, which is scarce in these regions, declining soil fertility over the long term (Pretty 2002). There is tremendous scope for increased organic agriculture, which currently accounts for only 2 per cent of global agriculture (0.37 million sq. km [including in conversion areas] of the total 50 million sq. km) (Willer 2011). In addition to environmental considerations, organic agriculture should also take social impacts (such as labour conditions) into account to provide truly sustainable means of food production.

In general, the environmental impacts of organic agriculture are low; however, in some specific cases they can be more than that of non-organic agriculture. For example, on sloping areas mechanical weeding compared to spraying with chemical weedicides could exacerbate erosion. Mulching with polyethylene sheets (permitted in organic agriculture to enhance soil moisture and quality) could be more harmful than the use of some chemical fertilizers. Flame weeders permitted in organic farming (using fire fuelled by liquid petroleum gas (LPG) to burn weeds) can be more energy and cost intensive than some chemical weedicides and less efficient in the control of perennial weeds (Wu and Sardo 2010: 52). These differences illustrate the importance of context specificity; in each specific case the actual environmental impacts could differ. However, if fossil-fuel-dependent machines and materials (such as plastics and flame weeders) are avoided, organic practices most often have lower impacts.

Organic agriculture practices and recommendations across Asia

Knowledge and labour intensiveness of organic agriculture. Although organic agriculture consumes less material and energy, it is labour and knowledge intensive. To overcome this challenge, training programmes and demonstrations are needed. In India the National Project on Organic Farming has organized over 4,500 training programmes and 6,750 demonstrations to train 270,000 farmers, create awareness, transfer technology and motivate farmers to adopt organic farming on more than 1.77 sq. km, over nine years since its inception in 2004 (ICCOA, 2013). This national project has created national and regional centres of organic farming to carry out research and dissemination activities. In Malaysia, studies have shown that livestock can lower labour costs by up to 50 per cent per hectare per year while also reducing weeding cost by 30 to 50 per cent and lowering chemical usage (Tayeb 2001; Ahmad 2001: 4). More such efforts could be replicated across Asia for the further dissemination of organic agriculture.

Programs to encourage urban and peri-urban kitchen gardens. Policies to promote organic agriculture could focus on the growing urban population who are keen to grow their own food in their gardens and on their rooftops. For example, in the Philippines programmes such as 'Food Always In The Home' (FAITH) (Lina 2014), Community Supported Agriculture (CSA) organizations such as Organic Manila and in India organizations such as *'Oota from your Thota'* (In Kannada *Oota* means food and *Thota* means garden) could be examined and replicated across Asia.

Convenient access to organic produce. Across Asia, the number of organic stores and farmers' markets have steadily increased making it easier for consumers to access organic produce. Restaurants that serve organic fare are also on the rise. While the numbers of such markets are increasing, the cost of organic produce remains a barrier for mass consumption, as detailed directly below. Promoting access to organic produce in stores and markets is recommended by organizing fairs and by providing financial incentives (such as tax rebate and subsidies). Additionally, an examination of the environmental trade-offs of eating out versus eating at home is needed.

High cost of organic produce. In general, the price of organic produce is usually 15 to 50 per cent higher than non-organic produce. This high price is attributed to high labour costs, low yields of organic farming, and most significantly to the subsidization of chemical fertilizers and pesticides used by non-organic farms. In India, the government spent over USD 12,640 million in just one year (2008–09) on chemical fertilizer subsidies; this is over 450 times the amount spent over nine years (2004–2012) on the National Project on Organic Farming (Sharma and Thaker, 2009). Serious efforts to subsidize organic agriculture rather than non-organic methods are needed. Currently, organic produce consumption in urban Asia is mostly accessible to the growing middle classes and elite consumers. While organic production for export can contribute to local livelihood development and national revenue, feeding the general population with organic produce remains a challenge. That being said, produce grown in small rural lots with traditional methods and low chemical inputs are most likely 'organic', though not certified as such, and are consumed by local populations.

Context-specific practices in the cities of Bengaluru, Kuala Lumpur, Metro Manila and Seoul

Organic agriculture in Bengaluru, India

One of the first expositions of organic agriculture (Howard 1940) was based on an analysis of environmental-friendly agricultural methods practised over centuries in India (Narayan 2005: vi). In the 1960s, the green revolution significantly changed agricultural practices across the country. In 2004, the Government of India initiated the National Project on Organic Farming to popularize organic methods once more. Although, in 2013 India accounted for the highest number of organic producers in the world (650,000 of 2 million worldwide) only 0.4 per cent of India's land under cultivation is managed using organic methods (Willer and Lernoud, 2015). Ways to increase organic agriculture need to take into account general challenges (as presented above) and more specific opportunities that each city/region within India provide.

Online delivery of cooked organic food. In Bengaluru, the silicon city of the east, there is a rapid rise in online access to home-delivered goods including groceries and food from restaurants, with specialized services for the delivery of cooked organic food becoming increasingly popular (Hariprasad 2015). While this increases access to organic produce, an analysis on the environmental trade-offs of online versus offline procurement would be recommended.

Resurgence of traditional grains. The health benefits of indigenous millets have revived recent interest in local millets such as *kodo* millet, barnyard millet and foxtail millet in south Indian cities such as Bengaluru and Chennai (Narayanan 2015; Devi et al. 2014: 1037). Increased production and consumption of organically grown millets is a favourable trend towards more organic agriculture.

Organic agriculture in Kuala Lumpur, Malaysia

Like in most other countries, Malaysia's organic agriculture movement stems from a local demand for organic produce and an awareness of the harm caused by agricultural chemicals (Christopher 2012). In 1986, from a one-hectare farm, the non-profit company Center for Environment, Technology and Development (CETDEM) first brought organic agriculture in its contemporary interpretation to Malaysia. Ever since, the demand for organic produce has grown but its supply is still a challenge. Currently, only 0.01 per cent of the country's agricultural land is under organic farming methods and Malaysia currently imports organic produce from Australia, the United States of America (USA) and New Zealand (Christopher 2012). To address this supply issue, the Third National Agriculture Policy (NAP3) 1998 identified organic agriculture as a market opportunity for Malaysian farmers (Ahmad, 2001). To help farmers certify their organic produce, certification is carried out by the Department of Agriculture's Skim Organik Malaysia. Lowering the cost and reach of certification would be one step forward, in terms of leveraging this policy effort.

Focus on major crops. Most studies on the environmental impact of agriculture in Malaysia are on palm oil cultivation (Tan et al. 2009: 420–427; Yusoff and Hansen 2007: 50–58). Further research is needed to understand the environmental impacts of the entire Malaysian agricultural sector, with a special focus on major crops and food products such as rice, pulses and meat.

High dependence on imported food. Most food products are cheaper to import from neighbouring countries than to produce in Malaysia (Murad et al. 2008, 609). Whether the same is true for organic produce is not known. As imported food has higher environmental impacts, a thorough analysis of the economic, geo-political, social and environmental factors related to this trend would require further investigation. Malaysia could benefit from investigating World Trade Organization mandates and its contribution to import dependency (Wade 2003).

Organic agriculture in Metro Manila, Philippines

Recent developments to support organic agriculture in the Philippines began as grassroots movements in the 1980s (Carating et al. 2010), leading to the passage of the Organic Agriculture Act of 2010. While some criticize the cost of the national certification scheme and continue to promote non-certified participatory guarantee systems, others see the Act as an opportunity to promote organic food production for local consumption and for export (Leuzinger et al. 2015). Currently only 0.8 per cent of the agricultural land in the Philippines is under organic agriculture (Willer and Lernoud, 2015). Further efforts are needed to promote organic food production and consumption.

Research on organic agriculture (especially rice). The International Rice Research Institute (IRRI) headquartered in the Philippines has been instrumental in conducting pioneering research to understand the environmental impacts of rice cultivation and methods to reduce these impacts (IRRI n.d.). Rice cultivation is one of the most significant sources of anthropogenic methane (Su et al. 2015). Research has shown several ways to reduce methane emissions from rice by replacing chemical fertilizers such as urea and ammonium sulphate with rice straw and green manure (Wassmann et al. 2000: 112). These practices could be further disseminated, in the Philippines and across the region. Although efforts are underway to produce organic rice, the price remains onerous for the average Filipino consumer.

Organic agriculture in Seoul, South Korea

The purpose of Korea's 2010 Management and Support for the Promotion of Eco-Friendly Agriculture/Fisheries and Organic Foods, also known as the New Organic Regulations (enacted by the Ministry of Agriculture, Food and Rural Affairs), is to increase the role of agriculture in preserving the environment (Ministry of Government Legislation 2009). This Act states the obligations, responsibilities, action plans and certification regulations for farmers, government and private organizations. In 2014 and 2015, South Korea signed agreements with the USA and the European Commission respectively for organic products certified in South Korea to be sold in these regions (Office of the United States Trade Representative 2014; European Commission 2015). Such systematic administrative support could result in an increase in the percentage of organic agricultural land from its current 1.1 per cent (Willer and Lernoud 2015).

National advocacy for organic agriculture. National efforts to spearhead South Korea's position in the global organic market, such as being the first Asian country to host the Organic World Congress (OWC) in 2012 (Paull 2012: 2), and opening the first museum dedicated to organic agriculture – *Namyangju* Organic Museum – are an important step. These efforts could be coupled with more local initiatives including training and education to raise production and consumption of organic food.

Data availability for LCA studies. Since 2009 the South Korean Rural Development Administration has established a Life Cycle Inventory (LCI) database for major crops such as rice, barley and beans. They have also committed to developing an LCI database for fifty items by 2020 (Yoon and Son 2012). Although, the domestic LCI database for seeds, seedlings and agrochemicals is only in its initial stages (Yoon and Son 2012), the amount of LCI information available in South Korea is greater than that available in most Asian countries. LCAs that have been conducted in South Korea for rice (Lim et al. 2010) and lettuce (Ryu and Kim 2010) could inform the framing of local policies that subsidize low-impact farming methods.

Impacts of food waste treatment: anaerobic digestion, composting, animal feed production, incineration and landfills

Food wasted during different stages of its life cycle, including wastage during agriculture, storage, transportation and post consumption, contributes significantly to environmental impacts due to:

1   Embedded environmental impacts that have already occurred in life stages prior to the stage at which the food is wasted;
2   Methane, nitrous oxide and leachate that is generated during its decomposition (Sonesson et al. 2005: 418).

The most effective strategy for reducing environmental impacts of food waste is to reduce the total amount of food waste generated at all stages of its life by understanding the reasons behind why wastage occurs and then by treating waste in ways that reduces noxious emissions. The two conventional means of managing waste – landfills and incineration – can have significant adverse impacts and result in the loss of valuable recyclable material such as glass, metal, paper and plastic. In addition, the enormous amount of waste generated by urban dwellers is making landfills unviable as they pollute ground water and soil, and increase the spread of disease. Cities therefore are investing in technologies to treat food waste after it has been segregated from recyclable, non-biodegradable materials (plastic, metal, glass) at source. These technologies include: anaerobic digestion with biogas production, conversion to animal feed (dry or wet), compost and vermi-compost production (Bernstad and Jansen 2011: 1879).

A review of life cycle assessment (LCA) impacts of food waste processing technologies establishes that incineration and landfills have the highest impacts; anaerobic digestion of food waste with biogas production has the least environmental impacts, followed by production of wet and dry animal feed, and compost (Figure 2.2) (Patterson et al. 2011: 7318; Bernstad and Jansen 2011: 1886; Khoo et al. 2010: 1373; Björklund et al. 1999: 53; Kim and Kim 2010: 4003). Environmental benefits of anaerobic digestion with biogas production depends on the environmental profiles of the fuels that biogas substitutes. If the biogas from food waste substitutes fossil fuels such as diesel and petrol, its impacts are much lower than if the biogas substitutes natural gas, or wood. In the United Kingdom (UK) and several other countries, anaerobic digestion is regarded as the most economic and technically appropriate method for treating organic waste as it simultaneously meets the targets for fossil CO_2 reduction, renewable energy generation and landfill diversion (Patterson et al. 2011: 7313).

Each of the treatment options represents specific challenges. For example, animal feed needs to be treated properly to avoid the spread of veterinary disease. Anaerobic digestion and compost production can release carcinogens into the soil and water. These carcinogens are from heavy metals that come from batteries, paint and other heavy metals that contaminate the feedstock (Patterson et al. 2011: 7318). Proper segregation of waste to avoid contamination of heavy metal containing substances and treatment of waste-derived animal feed can help overcome these challenges.

In order to devise a comprehensive city-wide waste management strategy, authorities could consider the environmental, social and economic trade-offs associated with various waste treatment options. Detailed information on all waste-related activities including collection, amounts of waste generated, composition of waste, mode of handling and treatment, energy recovery options, transport characteristics (vehicle details: type, fuel used, fuel efficiency, capacity, life span) can be fed into waste-LCA models to generate results useful to authorities and civil society (Gentil et al. 2010: 2637). These characteristics play a crucial role in deciding the waste management strategy for a city. For example, in Bengaluru 60 to 70 per cent of the municipal solid waste consists of organic (fermentable) waste (vegetable peels, leaves, bones from food waste) (Chanakya et al. 2009) whereas in Seoul this percentage is 23 per cent (Yoon and Jo 2002: 4) to 30 per cent (Yoo 2013), perhaps making anaerobic digestion more viable in Bengaluru.

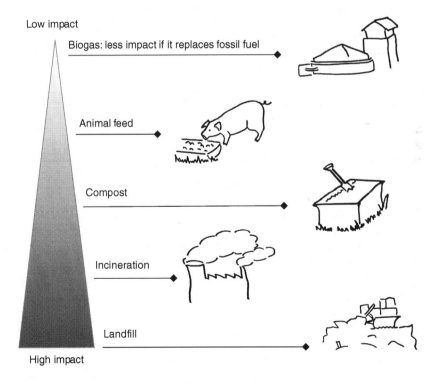

Figure 2.2 Low to high impacts of five food-waste processing technologies. Drawings by Megha Shenoy.

Anaerobic digestion can contribute to global warming due to leakage of methane (Berglund and Börjesson 2006: 476) (although there is no evidence of its link to fire hazards) and non-tilling of land after application of digestate (Møller et al. 2009: 3). Similarly, aerobic composting can emit GHGs if the compost is not properly aerated (such as with the assistance of earthworms) (Lundie and Peters 2005: 278).

Food waste treatment practices and recommendations across Asia

In general, recommendations are to treat segregated food waste for biogas extraction and/or compost production.

Segregation of waste at source. In most developing countries in Asia, there is either no policy on waste segregation at source or no enforcement of these policies. In Malaysia, the government is planning to enforce the Solid Waste Management and Public Cleansing Act 2007, which mandates source segregation starting from 1 September 2015. However, most people are still unaware of this law (Anon. 2015) and it is currently not enforced with any specific measures. In Bengaluru, although there is a law on waste segregation, it is estimated that only 30 per cent of waste is segregated at source, some of which is then mixed on its way to landfills (Dutta and Pallavi 2014). One of the primary roadblocks to enforcing segregation is an existing contractual system that pays transporters per ton of waste they collect and dispose. This is the case in Bengaluru and Metro Manila. These city-issued contracts could be reformulated to incentivize source segregation and recycling.

Incineration of mixed waste. Incineration of mixed waste, with or without energy recovery, is not recommended in developing countries for two reasons:

1. Food waste constitutes 60–70 per cent of the municipal solid waste (MSW) in most developing countries (Saeed et al. 2009: 2210; Chanakya et al. 2009). Due to the high moisture content of food waste – the calorific value of mixed waste is quite low. This low calorific value has led to the failure of several mixed waste incineration and waste-to-energy plants (Kasturirangan 2014: x, xiii).
2. Valuable recyclables such as paper, plastics and metals are lost due to incineration. These materials are the basis of the livelihood of millions of people in the informal recycling sector in these countries.

Despite these reasons, countries continue to opt for waste incineration. The Philippine government's National Solid Waste Management Commission (NSWMC) has drafted new rules to allow for waste-to-energy plants amid protests by environmental groups, and the Malaysian government has commissioned new mixed waste-to-energy plants (Ranada 2014; Neville 2010). In order to conserve recyclables and generate lower environmental impacts, an integrated approach that mandates segregation of waste at source followed by bio-methanation/composting of segregated biodegradable waste is recommended over incineration of mixed waste (Kasturirangan 2014: x; Shekdar 2009: 1443). Other technologies, such as the production of wet and dry animal feed, could also be examined and implemented.

Food waste as animal feed. In most Asian countries, the demand for meat is increasing (Delgado 2003: 3908S–3909S) (as discussed in Chapter 1). Some restaurants and canteens in Bengaluru give or sell their food waste to piggeries (Aravamuthu 2012). More research on these initiatives could be relevant, including on how to scale up efforts and ensure livestock feed safety.

Context-specific practices for cities of Bengaluru, Kuala Lumpur, Metro Manila and Seoul

Food waste treatment in Bengaluru, India

Reports on the amount of waste generated in Bengaluru vary hugely from 1700 MT/day to 4000 MT/day; this variation is partly due to unreliable reports from corrupt contractors who are paid per tonne of waste disposed (Ramachandra 2011; Chanakya et al. 2009). In 2012 the High Court of Karnataka ordered the city corporation (Bruhat Bengaluru Mahanagara Palike [BBMP]) to close operations at one of Bengaluru's biggest open landfills, following intense protest by residents in this area. BBMP then made it mandatory for citizens to segregate their waste into recyclables, biodegradables and biomedical fractions by amending the Karnataka Municipal Corporations (Amendment) Act 2013, which imposes fines on non-segregation and littering (Government of Karnataka, 2013). However, enforcement of this Act remains a challenge. Recommendations of this study follow from an understanding of the current status, local values and routines.

Societal value of leftovers. Traditionally, food in India is considered holy and thereby wastage of food is frowned upon. Leftover food is also considered to be of lower value (nutritionally and hygienically), thus not served to pregnant women (Choudhry 1997: 535), guests and male household members. Household members and domestic helpers in charge of cooking (in most cases women) make efforts to quantify the amount of food cooked per day so as to reduce leftovers and wastage. It would be important to encourage social practices, centred around the value of food, to lower its wastage.

Inconsistent power supply and effects on food storage and wastage. Most of India, even large cities, have inconsistent access to power supply, with outages experienced two to six hours per day, especially during the summer months. Therefore, most urban middle-class households do not stock food in the refrigerator, as is common in industrialized countries. This practice, and its correlation to food wastage, could be further examined.

Access to domestic help. Most urban middle-class households employ a domestic helper for household chores, with oftentimes a person charged with preparing fresh meals daily. Whether this practice results in better or worse management of household food stock could be further examined (Leray et al. in press; Karanth 2014).

Food waste treatment in Kuala Lumpur, Malaysia

Malaysia's Solid Waste Management and Public Cleansing Act 2007 focuses mainly on the construction and licensing of waste management facilities (Government of Malaysia 2007). As a result, currently 75 per cent of Malaysia's waste is landfilled, only 5 per cent is recycled whereas the remaining amount is illegally burned or dumped into water bodies (Abas and Wee 2014). This Act contains a general section on the 'reduction and recovery of controlled solid waste' but needs to elaborate on strategies that stem from local practices and values, such as segregation at source and levying a landfill tax.

Food waste treatment in Metro Manila, the Philippines

The Philippines followed in the footsteps of the USA with the passage of the National Solid Waste Management Act in 2000 after a series of national disasters, including a massive landslide of solid waste at the Payatas landfill in Metro Manila. Segregation at source is currently underway, with local governments responsible for reducing the amount of waste that ends up in landfills. The country also passed a Clean Air Act in 1999 that banned all waste incineration, a topic that continues to divide public opinion. In regards to reducing and better managing food waste, the following elements could be relevant.

High consumption of packaged food. Urban middle class households in Metro Manila purchase significant quantities of packaged and pre-prepared food (Burger Chakraborty et al. 2016). Packaged food in general has higher environmental impacts compared with fresh produce; yet packaging can play a role in lowering impacts by informing consumers about the products' environmental ratings, origin and agricultural practice (organic, greenhouse, non-organic) (Jungbluth et al. 2000: 138, 136).

Consumption of food waste. The poorest of the poor in Metro Manila purchase and consume leftover food that is scavenged from trash and sold as *pagpag*, a Tagalog term. The government's National Anti-Poverty Commission (NAPC) warns against eating this recycled food due to its disease burden and adverse effects on children's nutrition, and recommends that families apply for cash transfers to buy food instead. Despite this, poorer populations continue to derive their nutrition from food waste (Cabrera 2008). Society and the government could work together to find ways to get excess, uneaten food from the catering sector (restaurants, canteens and hotels) to people in need while also ensuring that the food adheres to health and safety standards. Organizations such as the Food Donation Connection, which have been involved in such initiatives in other countries (FDC 2014), could serve as models.

Food waste treatment in Seoul, South Korea

In 2007, South Korea's Wastes Control Act (of 1986) was amended to shift its focus away from maximizing waste treatment facilities and towards minimizing waste generation. It was also amended to redistribute the responsibility of managing waste among consumers, producers and the government rather than placing this responsibility solely with the government (Legislative Council Secretariat 2013). In 2008, the Act on Promotion of Saving and Recycling of Resources (of 2002) was also amended to promote extended producer responsibility by enforcing mandatory recycling rates for various products and packaging materials. South Korea's policies, as detailed below, can serve as model for other countries as they have taken relevant approaches such as:

Novel approach to penalizing food waste. Since 1997, the South Korean government has taken proactive policy decisions to reduce food waste generation including a ban on landfilling food waste from 2005 (Lee et al. 2007: 42). Since 2015, Seoul has introduced an innovative, high-tech system to reduce the generation of food waste. This system involves bins where the amount of food wasted by residents is weighed and a fee is charged depending on the amount of food wasted. In districts where this system has been on trial, an average household pays around Won 1300 per month (Euro 1) (Borowiec 2014). Although this amount may not appear significant, households and restaurants in these districts have reduced food waste by 30 per cent and 40 per cent respectively. This system is not seen as intrusive, as it stems from South Korea's culture of resourcefulness (Chrobog 2015). It could be more widely promoted across Asia, yet would depend on local cultures and how such laws would be enforced in different contexts.

LCA studies of waste management. An LCA study comparing the environmental impacts of landfills, incineration, compost and animal feed production in Seoul, found that landfills have high impacts to human toxicity and global warming; impacts of composting and animal feed production are from acidification, eutrophication and fresh water aquatic toxicity. Such studies could be used to inform policies on ways to further reduce the impacts of food waste treatment. Other countries could learn from these strategies, yet LCA data remains difficult and costly to access.

Conclusion

Food is one of the largest and oldest causes of land use and anthropogenic environmental impacts. It is also a vulnerable target of environmental hazards caused by climate change such as floods, droughts and changing rainfall patterns (Sachs 2014). These hazards and threats will create unpredictable and dire consequences not only for food security but also for food transportation, industrial food processing and waste treatment. An improved understanding of the relationship between food and its impacts across its entire life cycle can lead to more sustainable forms of food consumption and production.

The advent of the green revolution in the 1960s (Hazell 2009: 22) and the rise of consumerism in Asia have drastically changed the way we use energy, chemicals and water in the agricultural sector, and the way we consume and dispose materials. Identifying environmental impacts of food at the life stages of agriculture and food waste treatment using a review of LCA studies, along with an assessment of local practices and policies, can help uncover general trends that cities can use for sustainable development. This is especially significant for developing countries that lack LCA data. The review process is a low-cost option to identify alternatives with low environmental impacts. Authorities at the city level can use this information to implement technology alternatives with low impacts and examine policies that have worked in other contexts. Life Cycle Assessments (LCA) are gaining popularity in informing practices and policies that govern and manage multifaceted issues such as waste treatment and biofuel subsidies (Tasaki 2015: 236; Björklund and Finnveden 2007: 56).

This study also illustrates that while sustainable food production and consumption policies could lower impacts by carrying out LCA studies, it is also important to view solutions in a local context. Current strategies and policies rarely take into account practices that are rooted in local culture. They are usually based on a copy-paste paradigm that consists of redrafting policies that have been originally framed in developed countries. Additionally, these polices are carried out in a disconnected manner by departments and bureaucracies that are interconnected in the real world but are disjointed in their administrative duties. Policies could introduce a resource conservation approach across the entire food system using a framework that connects the different life stages of the food system: agriculture, wholesale and retail consumption, food waste generation, along with its treatment and management. A biophysical understanding of flows and impacts would need to be complemented by an assessment of local practices, to embed any solutions in a societal context.

Acknowledgements

I am extremely grateful to the editors of this book, Marlyne Sahakain, Suren Erkman and Czarina Saloma, for their invaluable support and input in substantially improving the content and readability of this chapter. Research for this chapter was financially supported by the Swiss Network for International Studies (SNIS) funded project '(Un)Sustainable Food Consumption Dynamics in South/Southeast Asia: Changing Patterns, Practices and Policies among New Consumers in India and the Philippines', coordinated by Suren Erkman, Shalini Randeria and Marlyne Sahakian.

Notes

1 A 'new consumer' is defined as an individual typically from a four-member household with a purchasing power of more than USD 10,000 per annum in 2000 (Myers and Kent 2004).
2 Throughout this chapter, I refer to 'practices' when discussing consumption taking place within culturally grounded social systems, and to 'patterns' when I discuss consumption as a biophysical activity with related impacts.
3 Sustainability, in this chapter, is defined as "the possibility that human and other forms of life will flourish on the planet forever" (Ehrenfeld 2004).

References

Abas, M. and S. Wee. 2014. 'The issues of policy implementation on solid waste management in Malaysia.' *Issues* 2(3): 12–17.

Ahmad, F. 2001. 'Sustainable agriculture system in Malaysia.' Regional Workshop on Integrated Plant Nutrition System (IPNS), Development in Rural Poverty Alleviation, United Nations Conference Complex, Bangkok, Thailand, 18–20 September.

Anon. 2015. 'Public still ignorant of waste separation rule beginning September 1.' *The Malaysian Insider*, 27 May.

Aravamuthu, K. 2012. 'Wasted food: Journey of that half eaten plate.' *The Alternative*, 19 April.

Badgley, C., J. Moghtader, E. Quintero, E. Zakem, M. Jahi Chappell, K. Avilés-Vázquez, A. Samulon and I. Perfecto. 2007. 'Organic agriculture and the global food supply.' *Renewable Agriculture and Food Systems* 22(02): 86–108.

Berglund, M. and P. Börjesson. 2006. 'Assessment of energy performance in the life-cycle of biogas production.' *Biomass and Bioenergy* 30(3): 254–266. doi: http://dx.doi.org/10.1016/j.biombioe.2005.11.011.

Bernstad, A. and J. la Cour Jansen. 2011. 'A life cycle approach to the management of household food waste – a Swedish full-scale case study.' *Waste Management* 31(8): 1879–1896. doi: http://dx.doi.org/10.1016/j.wasman.2011.02.026.

Björklund, A. and G. Finnveden. 2007. 'Life cycle assessment of a national policy proposal – the case of a Swedish waste incineration tax.' *Waste Management* 27(8): 1046–1058. doi: http://dx.doi.org/10.1016/j.wasman.2007.02.027.

Björklund, A., M. Dalemo and U. Sonesson. 1999. 'Evaluating a municipal waste management plan using ORWARE.' *Journal of Cleaner Production* 7(4): 271–280. doi: http://dx.doi.org/10.1016/S0959-6526(99)00086-4.

Borowiec, S. 2014. 'Food waste around the world: South Korea.' *The Guardian*, 27 March. www.theguardian.com/lifeandstyle/2014/mar/27/food-waste-around-world (accessed 7 June 2016).

Burger Chakraborty, L., Sahakian, M., Rani, U., Shenoy, M. and Erkman, S. 2016. 'Urban food consumption in Metro Manila: Inter-disciplinary approaches towards apprehending practices, patterns and impacts.' *Journal of Industrial Ecology*, 20(3): 559–570.

Cabrera, M. 2008. 'Filipino poor scavenge for recycled food to survive.' *Reuters*, 30 April. www.reuters.com/article/2008/04/30/us-philippines-food-scavengersid USSP32323420080430 (accessed May 2014).

Carating, R., M. Fernando, Y. Abrina and S. Tejada. 2010. 'State of organic agriculture in the Philippines: The Philippine country report.' *Workshop on ANSOFT of AFACI PAN-ASIAN Project*, 29–30 November. www.afaci.org/file/anboard2/PHILIPPINES(word).pdf (accessed December 2015).

Chanakya, H., T. Ramachandra and K. Shwetmala. 2009. 'Towards a sustainable waste management system for Bangalore.' 1st International Conference on Solid Waste Management and Exhibition on Municipal Services, Urban Development, Public Works, Icon SWM, Kolkata, India.

Choudhry, U. 1997. 'Traditional practices of women from India: Pregnancy, childbirth, and newborn care.' *Journal of Obstetric, Gynecologic, & Neonatal Nursing* 26(5): 533–539. doi: 10.1111/j.1552-6909.1997.tb02156.x.

Christopher, T. 2012. 'Organic agriculture and food in Malaysia.' http://christopherteh.com/blog/2012/02/organic-agriculture/ (accessed 23 December 2015).

Chrobog, K. 2015. 'In South Korea, an innovative push to cut back on food waste.' *Yale Environment 360*, 20 May.

Delgado, C. 2003. 'Rising consumption of meat and milk in developing countries has created a new food revolution.' *The Journal of Nutrition* 133(11): 3907S–3910S.

Devi, P., R. Vijayabharathi, S. Sathyabama, N. Gurusiddappa Malleshi and V. Brindha Priyadarisini. 2014. 'Health benefits of finger millet (*eleusine coracana l.*) polyphenols and dietary fiber: A review.' *Journal of Food Science and Technology* 51(6): 1021–1040. doi: 10.1007/s13197-011-0584-9.

Dutta, A. and A. Pallavi. 2014. 'Stink of lethargy: Lessons from two cities.' *Down to Earth*, 31 March.

Ecoinvent. 2015. 'Activity overview for Ecoinvent 3.2, Undefined.' www.ecoinvent.org/files/activity_overview_for_users_3.2_undefined.xlsx (accessed 15 December 2015).

Ehrenfeld, J. 2004. 'Searching for sustainability: No quick fix.' *Reflections* 5(8): 1–13.

European Commission. 2015. 'European Commission and South Korea agree to an equivalence arrangement on organic trade.' News on 27 January 2015. http://ec.europa.eu/agriculture/newsroom/188_en.htm (accessed 7 June 2016).

Evenson, R. and D. Gollin. 2003. 'Assessing the impact of the Green Revolution, 1960 to 2000.' *Science* 300(5620): 758–762. doi: 10.1126/science.1078710.

FDC. 2014. 'Food donation connection.' www.foodtodonate.com/Default.aspx (accessed 19 June 2015).

Gentil, E., A. Damgaard, M. Hauschild, G. Finnveden, O. Eriksson, S. Thorneloe, P. Ozge Kaplan, M. Barlaz, O. Muller, Y. Matsui, R. Ii and T. Christensen. 2010. 'Models for waste life cycle assessment: Review of technical assumptions.' *Waste Management* 30(12): 2636–2648. doi: http://dx.doi.org/10.1016/j.wasman.2010.06.004.

Gold, M. V. 1999. *Sustainable Agriculture: Definitions and Terms*. Beltsville, MD: U.S. Department of Agriculture.

Gomiero, T., D. Pimentel and M. Paoletti. 2011. 'Environmental impact of different agricultural management practices: Conventional vs. organic agriculture.' *Critical Reviews in Plant Sciences* 30(1–2): 95–124.

Government of Karnataka. 2013. 'Karnataka Municipal Corporations (Amendment) Act, 2013.' Edited by Karnataka State Legislature. Bangalore: Karnataka Gazette.

Government of Malaysia. 2007. 'Solid Waste and Public Cleansing Management Act 2007 (Act 672).' http://faolex.fao.org/docs/texts/mal74261.doc

Guinée, J., M. Gorrée, R. Heijungs, G. Huppes; R. Kleijn; A. de Koning, L. van Oers, A. Wegener Sleeswijk, S. Suh, H. Udo de Haes, H. de Bruijn, R. van Duin and M. Huijbregts. 2002. *Handbook on Life Cycle Assessment. Operational Guide to the ISO Standards. I: LCA in Perspective. II a: Guide. II b: Operational Annex. III: Scientific Background*. Dordrecht: Kluwer.

Hariprasad, U. 2015. 'Bye bye kitchen … Hello online organic food!' *The Alternative*, 30 May.
Hazell, P. 2009. *The Asian Green Revolution*, Vol. 911. Washington, DC: International Food Policy Research Institue.
Howard, A. 1940. *An Agricultural Testament*. New Delhi: Research Foundation for Science, Technology and Ecology.
Hughner, R., P. McDonagh, A. Prothero, C. Shultz and J. Stanton. 2007. 'Who are organic food consumers? A compilation and review of why people purchase organic food.' *Journal of Consumer Behaviour* 6(2–3): 94.
ICCOA. 2013. *About Organic Sector*. Bangalore: International Competence Centre for Organic Agriculture. www.iccoa.org/About_Organic_sector.php (accessed 10 May 2014).
IPCC (ed.). 2007. *Climate Change 2007: The Physical Science Basis. Contribution of Working Group I to the Fourth Assessment Report of the Intergovernmental Panel on Climate Change*. Edited by S. Solomon, D. Qin, M. Manning, Z. Chen, M. Marquis, K. Averyt, M. Tignor and H. Miller. Intergovernmental Panel on Climate Change. Cambridge, United Kingdom and New York, USA: Cambridge University Press.
IRRI. n.d. *International Rice Research Institute – Research*. http://irri.org/our-work/research (accessed 12 May 2014).
Jungbluth, N. 2000. 'Environmental consequences of food consumption: A modular life cycle assessment to evaluate product characteristics.' *The International Journal of Life Cycle Assessment* 5(3): 143–144. doi: 10.1007/bf02978610.
Jungbluth, N, O. Tietje, and R. Scholz. 2000. 'Food purchases: Impacts from the consumers' point of view investigated with a modular LCA.' *The International Journal of Life Cycle Assessment* 5(3): 134–142. doi: 10.1007/bf02978609.
Karanth, G. 2014. *Cooks and Relation to Food Waste*. Bengaluru (personal communication).
Kasturirangan, K. 2014. *Report of the Task Force on Waste to Energy (Volume I): In the Context of Integrated MSW Management*. Edited by Planning Commission. New Delhi: Government of India.
Khoo, H., T. Lim, and R. Tan. 2010. 'Food waste conversion options in Singapore: Environmental impacts based on an LCA perspective.' *Science of the Total Environment* 408(6): 1367–1373. doi: http://dx.doi.org/10.1016/j.scitotenv.2009.10.072.
Kim, M. and J. Kim. 2010. 'Comparison through a LCA evaluation analysis of food waste disposal options from the perspective of global warming and resource recovery.' *Science of the Total Environment* 408(19): 3998–4006. doi: 10.1016/j.scitotenv.2010.04.049.
Lee, S-H., K-I. Choi, M. Osako, and J-I. Dong. 2007. 'Evaluation of environmental burdens caused by changes of food waste management systems in Seoul, Korea.' *Science of the Total Environment* 387(1–3): 42–53. doi: 10.1016/j.scitotenv.2007.06.037.
Legislative Council Secretariat. 2013. 'South Korea's waste management policies.' www.legco.gov.hk/yr12-13/english/sec/library/1213inc04-e.pdf (accessed 7 June 2016).
Leray, L., M. Sahakian and S. Erkman. (In press). 'Understanding household food metabolism: Relating micro-level material flow analysis to consumption practices.' *Journal of Cleaner Production*.
Leuzinger, T., M. Sahakian and S. Erkman. (2015). 'Strength or weakness in diversity? Uncovering the multiple organic and local agriculture perceptions and practices in the Philippines.' *Food, Feeding, and Eating In and Out of Asia*. Copenhagen, Denmark: Asian Dynamics Initiative, University of Copenhagen.

Lim, S., C. Lee and S. Yang. 2010. 'environmental impact and external cost analysis by LCA: Conventional vs. organic rice.' http://agris.fao.org/agris-search/search.do?recordID=KR2011002951 (accessed 26 March 2015).

Lina J. Jr. 2014. 'Philippine agriculture must go organic.' *Manila Bulletin*, 3 March. http://www.mb.com.ph/philippine-agriculture-must-go-organic/

Lockeretz, W., G. Shearer, and D. Kohl. 1981. 'Organic farming in the corn belt.' *Science* 211: 540–546.

Lundie, S. and G. Peters. 2005. 'Life cycle assessment of food waste management options.' *Journal of Cleaner Production* 13(3): 275-286. doi: 10.1016/j.jclepro.2004.02.020.

Ministry of Government Legislation. 2009. 'Environmentally-friendly Agriculture Fosterage Act.' www.moleg.go.kr/english/korLawEng?pstSeq=54780 (accessed 24 December 2015).

Møller, J., A. Boldrin, and T. Christensen. 2009. 'Anaerobic digestion and digestate use: Accounting of greenhouse gases and global warming contribution.' *Waste Management & Research* 27(8): 813–824.

Murad, W., N. Hashim Nik Mustapha and C. Siwar. 2008. 'Review of Malaysian agricultural policies with regards to sustainability.' *American Journal of Environmental Sciences* 4(6): 608–614.

Myers, N. and J. Kent. 2004. *The New Consumers: The Influence of Affluence on the Environment*. Washington, Covela, London: Island Press.

Narayanan, C. 2015. 'Comeback cereal.' *Business Today*, 10 May.

Narayanan, S. 2005. 'Organic farming in India: Relevance, problems and constraints.' Occasional Paper 38, Department of Economic Analysis and Research and National Bank for Agriculture and Rural Development, Mumbai. www.nabard.org/pdf/OC%2038.pdf (accessed 7 June 2016).

Neville, A. 2010. 'Top plant: Kajang waste-to-energy plant, Semenyih, Malaysia.' *Power: Business & Technology for the Global Generation Industry since 1882*. Houston, TX: Electric Power, USA.

Office of the United States Trade Representative. 2014. 'United States and Korea streamline organic trade.' https://ustr.gov/about-us/policy-offices/press-office/press-releases/2014/July/United-States-and-Korea-Streamline-Organic-Trade (accessed 7 June 2016).

Patterson, T., S. Esteves, R. Dinsdale and A. Guwy. 2011. 'Life cycle assessment of biogas infrastructure options on a regional scale.' *Bioresource Technology* 102(15): 7313–7323. doi: 10.1016/j.biortech.2011.04.063.

Paull, J. 2012. 'Asian spring for organic agriculture: Korea takes a lead.' *Journal of Organic Systems* 7(1): 2–4.

Pimentel, D., P. Hepperly, J. Hanson, D. Douds, and R. Seidel. 2005. 'Environmental, energetic, and economic comparisons of organic and conventional farming systems.' *BioScience* 55: 573–582.

Pretty, J. 2002. 'Lessons from Certified and Non Certified Organic Projects in Developing Countries.' In *Organic Agriculture, Environment and Food Security*, edited by N. El-Hage Scialabba and C. Hattam, pp. 139–162. Rome, Italy: FAO.

Pretty, J. and R. Hine. 2001. *Reducing Food Poverty with Sustainable Agriculture: A Summary of New Evidence*. SAFE Research Project. Wivenhoe Park, UK: University of Essex.

Ramachandra, T. V. 2011. 'Integrated management of municipal solid waste.' In *Environmental Security: Human and Animal Health*, pp. 465–484. Lucknow, India: IDBC Publishers.

Ramesh, P., M. Singh and A. Rao. 2005. 'Organic farming: Its relevance to the Indian context.' *Current Science* 88(4): 561–568.

Ranada, P. 2014. 'Gov't to draft rules on waste-to-energy burning amid protests.' *Rappler*, 12 August. www.rappler.com/science-nature/environment/65996-guidelines-waste-energy-incineration# (accessed 7 June 2016).

Rao, I. 1999. 'Soil and environmental pollution – a threat to sustainable agriculture.' *Journal of the Indian Society of Soil Science* 47: 611–633.

Refsgaard, K., N. Halberg, and E. Kristensen. 1998. 'Energy utilization in crop and dairy production in organic and conventional livestock production systems.' *Agricultural Systems* 57: 599–630.

Ryu, J. and K. Kim. 2010. 'Application of LCA methodology on lettuce cropping systems in protected cultivation.' http://agris.fao.org/agris-search/search.do?recordID=KR2011002636 (accessed 26 March 2015).

Sachs, J. 2014. *What is Sustainable Development?* Course book chapters available on the free online course, 'Age of Sustainable Development' on Coursera.

Saeed, M., M. Hassan and M. Abdul Mujeebu. 2009. 'Assessment of municipal solid waste generation and recyclable materials potential in Kuala Lumpur, Malaysia.' *Waste Management* 29(7): 2209–2213. doi: 10.1016/j.wasman.2009.02.017

Safaei, A. 2015. 'Country specific LCIs in the Ecoinvent database.' Consultative meeting on readiness for creating Indian LCA database at Federation of Indian Chambers of Commerce and Industry, New Delhi on 16 September 2015.

Sharma, V. and H. Thaker. 2009. 'Fertilizer subsidy in India: Who are the beneficiaries?' *White Paper No. 2009-07-01*. Ahmedabad, India: Indian Institute of Management. www.iimahd.ernet.in/publications/data/2009-07-01Sharma.pdf (accessed 7 June 2016).

Shekdar, A. 2009. 'Sustainable solid waste management: An integrated approach for Asian countries.' *Waste Management* 29(4): 1438–1448. doi: 10.1016/j.wasman.2008.08.025

Sonesson, U., B. Mattsson, T. Nybrant and T. Ohlsson. 2005. 'Industrial processing versus home cooking: An environmental comparison between three ways to prepare a meal.' *AMBIO: A Journal of the Human Environment* 34(4): 414–421. doi: 10.1579/0044-7447-34.4.414.

Stolze, M., A. Piorr, A. Haring and S. Dabbert. 2000. 'The environmental impact of organic farming in Europe.' In *Organic Farming in Europe: Economics and Policy*, edited by S. Dabbert, N. Lampkin, J. Michelsen, H. Nieberg and R. Zanoli. Hohenheim, Germany: University of Hohenheim.

Su, J., C. Hu., X. Yan, Y. Jin, Z. Chen, Q. Guan, Y. Wang, D. Zhong, C. Jansson, F. Wang, A. Schnurer and C. Sun. 2015. 'Expression of barley SUSIBA2 transcription factor yields high-starch low-methane rice.' *Nature* 523(7562): 602–606.

Tan, K., K. Lee, A. Mohamed and S. Bhatia. 2009. 'Palm oil: Addressing issues and towards sustainable development.' *Renewable and Sustainable Energy Reviews* 13(2): 420–427.

Tasaki, T. 2015. 'Environmental subsidies and life cycle assessment.' In *Environmental Subsidies To Consumers: How Did They Work In The Japanese Market?*, edited by Shigeru Matsumoto, p. 236. New York: Routledge.

Tayeb, D. 2001. 'Integrated farming system for the sustenance of Malaysian oil palm industry.' Malaysian Soil Science Society Seminar, Kota Kinabalu, Sabah, Malaysia.

Wade, R. 2003. 'What strategies are viable for developing countries today? The World Trade Organization and the shrinking of "development space".' *Review of International Political Economy* 10(4): 621–644.

Wardenaar, T., T. Ruijven, A. Beltran, K. Vad, J. Guinée and R. Heijungs. 2012. 'Differences between LCA for analysis and LCA for policy: A case study on the consequences of allocation choices in bio-energy policies.' *The International Journal of Life Cycle Assessment* 17(8): 1059– 1067. doi: 10.1007/s11367-012-0431-x.

Wassmann, R., L. Buendia, R. Lantin, C. Bueno, L. Lubigan, A. Umali, N. Nocon, A. Javellana, and H. Neue. 2000. 'Mechanisms of crop management impact on methane emissions from rice fields in Los Baños, Philippines.' *Nutrient Cycling in Agroecosystems* 58 (1–3): 107–119. doi: 10.1023/a:1009838401699.

Willer, H. 2011. 'The world of organic agriculture 2011: Summary.' In *The World of Organic Agriculture: Statistics and Emerging Trends*, edited by H. Willer and L. Kilcher, pp. 26–32. Bonn, Germany: IFOAM.

Willer, H. and J. Lernoud (eds). 2015. 'The world of organic agriculture: Statistics and emerging trends 2015.' FibL-IFOAM Report. Research Institute of Organic Agriculture (FiBL), Frick, and IFOAM-Organics International, Bonn.

Wu, J. and V. Sardo. 2010. 'Sustainable versus organic agriculture.' In *Sociology, Organic Farming, Climate Change and Soil Science*, edited by E. Lichtfouse, pp. 41–76. Dordrecht: Springer.

Yoo, K. 2013. '2012–2021 Seoul Waste Management Plan.' First International Waste Working Group Asian Regional Branch (IWWG-ARB) Symposium, Hokkaido University, Japan, 18–21 March.

Yoon, E. and S. Jo. 2002. 'Municipal solid waste management in Tokyo and Seoul.' Workshop of IGES/APN Mega-City Project, Rihga Royal Hotel Kokura, Kitakyushu, Japan, 23–25 January.

Yoon, S. and B. Son. 2012. 'Life cycle assessment (LCA) for calculation of the carbon emission amount of organic farming material – with emphasis on hardwood charcoal, grass liquid and microbial agents.' http://agris.fao.org/agris-search/search.do?recordID=KR2013001905 (accessed 26 March 2015).

Yusoff, S. and S. Hansen. 2007. 'Feasibility study of performing a life cycle assessment on crude palm oil production in Malaysia.' *The International Journal of Life Cycle Assessment* 12(1): 50–58.

Part II
Food practices and cultures

3 Emerging consumerism and eating out in Ho Chi Minh City, Vietnam
The social embeddedness of food sharing

Judith Ehlert

The Vietnamese proverb 'Enough food and warm clothing; delicious food and beautiful clothing'[1] brings to mind the overall socio-economic transformation experienced in Vietnam since the Indochina Wars and the opening up of its economy in the mid-1980s. Whereas people relate the first part of the saying to the hardship of fulfilling basic needs during and in the aftermath of the war-torn decades, the second – 'delicious food and beautiful clothing' – can be read as representative of a societal longing for an improved quality of life, and the growing relevance of consumerism in the context of the country's economic development. Accordingly, food and fashion have become expressions of taste and well-being and, fuelled by rapid economic growth and the modernisation of the country's food system, as well as the resurgence of the urban middle classes and their appetite for 'foreign' food, Ho Chi Minh City (HCMC) has become a hotspot for eating out in Vietnam. More than just the financial capacity to purchase 'delicious food', eating at a diverse range of gastronomic locations – from indoor, air-conditioned, upmarket cafés to fast food outlets – demonstrates a person's knowledge and command of up-to-date lifestyle trends regarding how to socialise and consume 'properly' (see Welch Drummond 2012) and, by extension, offers a means of expressing a modern, urban lifestyle and 'middle-classness' (Bélanger et al. 2012; Bitter-Suermann 2014; Earl 2014).

This chapter approaches emerging urban food consumerism and the eating out trend from the perspective of sharing food as a social practice (note that several chapters in this volume pick up this trend, considering eating out from Manila in Chapter 4 to Melbourne in Chapter 6). By drawing on and redefining Bourdieu's concept of 'habitus', it sheds light on two different but interrelated phenomena: the structural context of food consumption and its social embeddedness.

After the presentation of the conceptual framework and methodological approach, this chapter begins with an overview of the key features of Vietnam's food system's modernisation and the consumption landscape in HCMC. After a brief note on macro-level changes, it continues with a consideration of the actual social practices of sharing food outside one's home and how those practices are changing in terms of gendered and generational dimensions of identity construction via food consumption. Vietnam's rapid economic integration and growing consumerism has led to increasing food sharing options, affective consumption of global consumer culture, and diverse forms of social distinction and belonging. At the same time, overall growing social inequality (see, for example, King et al. 2008; Taylor 2004) can also be observed in the food sector. The practices of food sharing will thus be discussed alongside the practices and structures of food waste, with the aim of drawing attention to the need for further sociological and anthropological understanding of the social inequalities embedded in the food consumption dynamics of one of Asia's mega-cities. Despite Vietnam's priority in terms of economic growth and development, the social implications of the growing eating out trend do not yet feature in contemporary debates in Vietnam and remain neglected in terms of scholarly inquiry into Asia's 'new consumers' and the country's development pathway.

Conceptual framework and methodological approach

According to Beardsworth and Keil (1997: 100–122), eating out options in general emerge in accordance with changes in the social organisation of work, urbanisation, and processes of social mobility and migration. Against such a background of structural transformation, eating is turned into a commercial transaction and thereby removed from its formerly exclusive framework of family- and kin-based social obligation and reciprocity. Nevertheless, eating together and food in general continue to constitute important markers of social solidarity: 'Sharing food is held to signify "togetherness", an equivalence among a group that defines and reaffirms insiders as socially similar' (Beardsworth and Keil 1997: 101; referring to Mennell et al. 1992). How this 'togetherness' is performed in contemporary eating-out settings will be the focus.

Launched in 1986, an economic reform programme called *đổi mới* facilitated Vietnam's recent history of (re-emerging) consumerism, framing the material context, the institutional setting and the actual practice of consumption by newly 'discovered' consumers, which had been quashed under communism (see below). Consumption in this chapter is conceptualised as integrating the structural context of consumption as well as the daily practices, routines and meanings ascribed to consumption by consuming actors. The latter relate to the affective-symbolic dimension, which has to do with ingesting and embodying (see also Coles 2013: 255) what is presumed to constitute, for example, 'food modernity' or 'global consumer culture' (on consumption and social practice theory see for example Warde 2014; for Vietnam see Hansen 2015).

The study presented considers both the structural context of the food system and the material availability of goods in Vietnam, as well as the everyday social practices of eating out. According to this understanding, emerging consumer markets and eating-out settings change the social landscape of eating (Julier 2013). In return, eating as a social practice has an influence on how consumer markets and culinary landscapes are shaped. By bringing together structure and agency in the tradition of Bourdieu (1977) and Giddens (1984), this chapter aims to understand how the modernisation of Vietnam's food system (in terms of provisioning products, supermarkets and gastronomic infrastructures, among other dimensions) provides the material conditions in which the socio-cultural practices of eating out take place. At the same time, it addresses how various gendered and generational food consumption norms are (re-)produced, negotiated and challenged through the everyday practices of sharing food, and vice versa. Furthermore, how the status-enhancing role of food and the practice of over-ordering affect the establishment of a modernised food system and its production-driven externalities will also be considered.

Bourdieu's concept of 'habitus' (Bourdieu 1984) helps to explain the saying 'People are what they eat', as it reveals the structural (class) and individual dispositions of certain consumption practices. Under the umbrella of 'taste', Bourdieu analyses, for example, the different values people ascribe to certain foods while tabooing others, and the manner and settings in which food is enjoyed. Accordingly, people's habitus of eating is more or less anchored in the existing structure of class relations and materialises empirically in the lifestyles of certain social groups and their corresponding practices of eating. However, for the purposes of this chapter, the habitus concept and its rather static and structural interlinkage between class and lifestyle falls short of fully capturing the strongly processual and fluid dimension of 'doing middle-classness', social differentiation and boundary-drawing observable in Vietnam. Hence, what is meant by 'class' in Vietnam in the first place constitutes a conflict-laden terrain of negotiation between the country's socialist legacy and the ideal of communist class structure, and current neoliberal economic forces fostering individualisation and aggravating social inequalities (see Schwenkel and Leshkowich 2012). The middle class in Vietnam cannot be identified by rigorous and clearly defined structural indicators (see for example King et al. 2008; Nguyen-Marshall et al. 2012). Suggested here, therefore, is the perspective of 'doing middle-classness', which focuses on the *social practices* by which middle-class identity is (re-)constructed by the constant testing and manoeuvring of, for example, gendered and generational boundaries. The meaning of social class and lifestyle in such a dynamic development context as that of Vietnam remains in flux. To begin with, this contribution centres on the gendered spaces and inter-generational aspects of 'eating away from home' in the vibrant urban space of HCMC.

The research on which this chapter is based is mainly drawn from fieldwork carried out in HCMC over a total of seven months between 2013 and 2015 in the context of a broader research project on the body politics of eating and the transformation of food culture in Vietnam.[2] Observation was an important aspect of this fieldwork. The rapidly diversifying gastronomic scene served as a methodological entry point through which to approach broader societal changes, for example gendered and generational relationships as manifested in the social practices of food sharing. Practically, this meant spending time in different eating and drinking venues such as fast food outlets and coffee shops. Based on theoretical sampling and the decision to put the (re-)emerging urban middle class under the microscope, these observations focused on upscale eateries and coffee shops that can be classified as geared towards entertainment, leisure and lifestyle purposes rather than the daily reproductive necessity of eating. As a certain affiliation of youth culture with fast food outlets and coffee shops emerged from these observations, two group discussions with female and male university students were conducted. These were then contrasted with oral history interviews in order to elaborate the generational and gender dimension involved in the transformation of food consumption contexts and practices. Furthermore, interviews were conducted with scholars of Vietnamese food history and education, a market research institute staff member, and a manager and waiter at two upmarket restaurants characterised by higher-priced menus, a business clientele and prime locations. The interviews focused on overall food culture changes and consumption trends. Certain dimensions of food waste as an aspect of food sharing implicitly emerged in the data collection process and were taken up in the theoretical sampling process through an initial exploration of Vietnamese (online) newspapers and social media, as well as a narrative interview with the founder of an urban food bank initiative in Hanoi.

Manoeuvring foodscapes and consumption in HCMC

The last few decades have been characterised by rapid economic growth in Vietnam. Initiated in 1986, *đổi mới* and its portfolio of economic reforms supporting privatisation, market liberalisation and the decollectivisation of agriculture, facilitated the shift from a planned to a market economy (Beresford 2001) and paved the way to global economic integration, as demonstrated by the country's accession to the World Trade Organization in 2007. This economic pathway has been accompanied by a dynamic transformation of the overall food system, which has transformed Vietnam's past experience of severe food scarcity into one of growing diversification of, yet unequal access to, food consumption options in urban centres.

Whereas the Food and Agricultural Organization, in its report *Food Consumption and Nutritional Status in Viet Nam* (1990), describes the country in 1990 as far from food secure, and the nutritional status of the population as unsatisfactory, the picture nearly three decades after *đổi mới* looks quite different. Export-orientated agricultural reforms stimulated a production surplus and, in general, facilitated unprecedented economic development (Beresford 2001; Bui Van Hung 2004). Besides growing global market integration, industrialisation and rapid urbanisation have played their parts in the modernisation of the food system and, of most relevance here, in the stimulation of the consumption patterns of the (re-)emerging urban middle classes.

HCMC makes for a particularly interesting case study as it is the most rapidly developing urban centre in Vietnam and in the Southeast Asian region (Earl 2014). After the French colonial defeat of 1954 and the separation of the country into the communist Democratic Republic of Vietnam in the north and the US-backed capitalist Republic of Vietnam in the south, the southern government in Saigon promoted urbanisation and attracted and produced a middle class engaged in the wartime capitalist economy. Conversely, Hanoi, as capital of the communist north, experienced a process of deurbanisation and ruralisation. When North and South Vietnam were reunited as the current Socialist Republic of Vietnam and Saigon was renamed Ho Chi Minh City in 1975, the southern urban middle classes at first experienced rapid downward social mobility in the name of communist egalitarianism (Earl 2014). Against this background, however, đổi mới provided the nurturing ground for the (re-) construction of HCMC's middle classes. Rural–urban migration (especially female) was stimulated by education and employment opportunities in the city, which eventually translated into new gains in assets, resources and social position.[3] Today, these forms of capital are invested in ever-growing consumption options. Food and eating out have become prominent markers of consumption for the performance of middle-classness (Earl 2014). In contrast to the Vietnamese middle class elite under colonialism, for which 'a[n exclusive] taste for new foods and culinary innovation was a way for people to highlight their cultural sophistication and social status' (Peters 2012: 43), food consumption today is a more mainstream way of performing middle-classness.

Due to integration into the global economic system, HCMC's consumption options have multiplied with the mushrooming of supermarkets. Vietnam has been experiencing a so-called 'third wave' of global supermarket shares since the late 1990s and early 2000s (Mergenthaler 2008), and increasingly opening up to consumers with varying degrees of purchasing power (Cadilhon *et al.* 2006; see also Figuié and Bricas 2010). Vietnam's retail sector ranks among the top bracket of countries receiving foreign direct investment (Mergenthaler 2008) and foreign brands find their way onto the Vietnamese market via supermarket supply chains. According to a female representative of a HCMC market research institute conducting panel studies on consumption behaviour in Vietnam, pre-packaged food today constitutes a huge segment of sales and part of a household's expenditure. Being a white-collar worker and mother herself, the market researcher explained the time pressure on working mothers:

> The middle class especially likes to adopt a Western lifestyle and food. … We also have food from other countries, no longer only Vietnamese food, so that is why we adopt the new products. And also because we have less time as we are busier with work. That is why some consumers now adopt more ready-to-eat products and semi-processed products as well. So they buy and just need time to heat it in the microwave.
> (Interview, HCMC, 21 September 2015)

Besides the development of supermarkets, a similar investment boom that gradually followed market liberalisation can be observed in the gastronomic sector and its highly diversified range of indoor restaurants and coffee shops. Between 2008 and 2012, for example, the fast food sector in Vietnam showed a compound annual growth rate of about 17 per cent (about 14 per cent for cafés/bars) which is forecast to reach about 18 per cent (remaining at about

14 per cent for cafés/bars) between 2013 and 2017 (Agriculture and Agri-Food Canada 2014: 2, 5). Shopping malls increasingly integrate fast food venues with entertainment and shopping. On the subject of shopping malls, the representative of the market research company explained:

> It is not only about the store for us to buy things. It is also a better place for entertainment for the whole family. So we bring kids there. We go with the whole family, we can go there to some fast food restaurant, just located in some supermarket. So it's kind of an entertainment place as well.
> (Interview, HCMC, 21 September 2015)

This is consistent with personal participant observation – that supermarkets and shopping malls are especially packed on the weekends and have dozens of product promotions that strongly attract customers. Through diverse media channels, food retail and the gastronomic business have become symbolically omnipresent in people's everyday lives. Product advertisements are placed on TV shows and oversized street banners. A plethora of women's magazines discuss the latest food trends, diets and the city's culinary hotspots (see also Drummond 2004). En vogue restaurants and foodies' insider tips are communicated to friends and peers via social media networks, as one female informant in her mid-twenties explained. She selects the places to go through social media, from an otherwise overwhelming and very fast-changing gastronomic landscape (interview, HCMC, 13 October 2015).

Whereas opulent, conspicuous consumption and 'imperialist' foreign goods once contradicted communist egalitarian ideology, the government and the economy perceive the Vietnamese consumer as an important actor following the initiatives of market liberalisation (Figuié and Moustier 2009), bringing the communist ideal of economic equality gradually under pressure. Against this structural context, newly 'discovered' consumers aim to experience the culinarily exceptional, and search for new modes of social distinction and belonging. This ongoing process of identity construction by the consumer is quite aggressively pursued by the marketing strategies of, for example, the soft drink sector, which is also booming. For instance, in 2014 and 2015 consumers were extremely excited about Coca Cola cans customised with personal names or a characteristic such as 'shy' or 'good looking' (interview market researcher, HCMC, 21 September 2015). This hype can be read as a practice of symbolically embodying and ingesting individuality.

In general, consumption options in HCMC are diverse. Food consumption in Vietnam, however, has only recently been taken up by scholars, and only as one of the many indicators of middle-class lifestyles (Earl 2014). To fill this gap, this chapter focuses on food consumption understood as a constitutive element in the ongoing process of identity construction in Vietnam. Consumers find themselves navigating the urban eating-out landscape as the material context of food diversifies and is discursively backed by marketing and advertisement strategies promoting modern consumption. The question thus arises: How do consumers actually appropriate this structural context into their own space? This leads from the macro-level context to the everyday social practices of food consumption, detailed as follows.

Eating out: changing social practices of food sharing

Food sharing in Vietnam is deeply embedded in family, kin and ancestral relations and incorporated in a dense script of having food together as a form of commensality (Avieli 2012). Norms of seniority, gender and hospitality, for instance, frame the social practices of sharing food. The growing trend towards eating and drinking out presented above, however, implies a shift from home-based commensality towards (or at least its complementation by) sharing food outside one's home with peer groups or co-workers, and eating on one's own alongside, but not with, strangers.

When one of the major multinational corporations, following other well-established foreign fast food chains, opened its first fast food restaurant in HCMC in 2014, customers – the young and the old as well as whole families – queued for hours to be among the first to experience this new gustatory and social event (BBC News, 8 February 2014).[4] In a group discussion about foreign fast food restaurants, male and female university students explained that the main reason to visit was not necessarily the food itself. One of the major reasons was that these places have air-conditioning and huge glass fronts. The students like to be seen through these large glass windows by others from the outside, thus displaying and communicating modernity to and social distinction from the world outside the restaurant (group discussion students, HCMC, 28 March 2013). The motivations of customer groups, especially professionals and white-collar workers, young couples and teenagers, as well as families curious about the culinarily 'foreign', have been documented in, for example, the case of Western-style fast food restaurants opening in late 1980s Beijing, offering what was at that time an unknown food experience contrasting with the common socialist canteens (Yan 2008). Not unlike the students in HCMC, most of the customers Yan interviewed reported that they did not like the taste of the imported fast food in particular but enjoyed the style of eating and the experience of being there: 'Most customers spent hours talking to each other and gazing out the huge glass windows onto busy commercial streets – and feeling more sophisticated than the people who passed by' (Yan 2008: 87).

Obviously, eating out constitutes a social practice that goes beyond socialising among peers. It also serves to delineate one's belonging to one group rather than another through the demonstration of a command of dining etiquette and social skills. In the case of Vietnam, the studies of, for example, Bitter-Suermann 2014, Welch Drummond 2012, Earl 2014 and Higgins 2008 similarly identify eating out and spending time in gastronomic locations as an expression of social status aspiration; it is the demonstrated capacity to align oneself with and adapt to what is perceived as modern global consumer culture, as well as to locally embed it (see Bitter-Suermann 2014: 38).

The following therefore looks into how this connection with and adoption of global consumer culture actually unfolds in the concrete social practices of drinking and eating out in HCMC, particularly in terms of gendered and inter-generational relations. In general, by sharing food and drinks outside the home, actors 'engage in [new] forms of sociability, delineating lines of intimacy and distance' (Julier 2013: 339). The argument proposed here is that this delineation is not a mere demarcation in terms of Bourdieu's class structure. Instead of applying the habitus concept as enshrined in more or less fixed class dispositions, the following starts from the very practices of creating those lines of intimacy and distance. As will be shown, gender and generational dimensions as well as social status are at play in the construction and negotiation of distinction and belonging.

Gendered spaces of drinking and eating out

Whereas coffee shops and beer halls used to be predominantly frequented by men, Welch Drummond (2012: 88) observes a rise since *đổi mới* in the number of public spaces in Hanoi that facilitate socialising among women, thus increasing their visibility. As documented above, the landscape and popularity of upmarket cafés and fast food outlets that Welch Drummond describes in Hanoi (2012: 88) can also be found in the indoor, air-conditioned locations of contemporary HCMC. The following narratives of two older female informants in their late fifties and sixties, Anh[5] and Ngoc, recall the restrictions put on women in the past in terms of spending leisure time and socialising in public. Their oral history perspective complements the younger generation's perception of the way being female or male continues to structure food-sharing practices.

Anh, during an interview in a coffee shop, told me that many years ago it would not have been possible for her to meet in such a location. Referring to Confucian legacies and pre-*đổi mới* times, she recalled that women never went out and spent time in coffee shops but simply stayed at home. She illustrated this through the example of her mother:

> My mother, she never goes out – even now. She is eighty-five years old and when she got married she was eighteen. Whenever I visit my parents, I always see her in the kitchen. And she never goes out without her husband – my father. But my father can go out anywhere. He only takes her out about five or six times a year. When friends come to visit her at home, while talking she always has to do something – some cleaning, some vegetables, some cooking. No sitting and talking [only] – no!
> (Interview, Anh, HCMC, 5 October 2015)

Anh's personal account echoes that of a Vietnamese food historian discussing the influence of Confucianism on the gendered social norms around eating and drinking in Vietnamese society prior to the communist egalitarian ideal after 1975. According to Vietnam's Confucian tradition, men and women in public ate separately and, when receiving guests, the woman did not join the feast: 'In the family, it was never the case that women had a meal with the guests; she just said a word of greeting, and [was responsible for] cooking after that' (interview, food historian, HCMC, 29 August 2014).[6]

Ngoc, the other female informant in her sixties, recalled her own upbringing and the food and drinking socialisation she experienced from childhood until she got married. As a girl she was told that she must not drink coffee in public as this was the behaviour of an 'unserious girl'. When she later went out with her husband and his male friends she had to order orange juice in order not to receive 'unfriendly looks' from the latter. In the same vein, men used to refrain from having, for example, orange juice in public, as this was highly associated with feminine drinking behaviour and did not comply with the 'correct' behaviour of men enjoying black coffee and cigarettes in coffee shops (interview, Ngoc, HCMC, 29 August 2014; see also interview, Anh, HCMC, 5 October 2015).

Comparing Ngoc's gendered socialisation with the conditions of the younger generation, those gendered norms of femaleness and maleness associated with certain drinking habits in the private and public spheres are gradually changing. Some of her female grandchildren and nieces actually like the taste of black coffee but are still taught that Vietnamese women who like black coffee are generally considered 'manly'. Girls are still educated to refrain from drinking black coffee in public. Based on my own observations in such coffee houses, they drink fruit juices and smoothies instead. Some would order the typical *café sữa đá* – iced milk coffee, consisting of coffee, ice and sweet condensed milk to make the bitter coffee sweeter, milder, softer and more 'womanly', thus embodying femininity. The popular *café sữa đá* works as a door-opener to the formerly male-dominated public café culture in as much as it merges feminine as well as masculine associated qualities – sweetness and bitterness. In this way, formerly gendered public spaces seem to have become gradually more and more porous and reorganised since *đổi mới*. However, another group discussion conducted with female and male university students in HCMC showed how Confucian principles concerning what and how to eat 'correctly' depending on one's being a man or a woman are still perceived as somehow obligatory.

Complementing the accounts of the older female informants, the younger female generation perceived certain female-ascribed characteristics such as frugality and empathy as being expected of them when sharing food with others (see also Avieli 2012). Whereas men can 'make themselves feel at home' at the table, young women reported being concerned with eating more moderately and gently (group discussion students, HCMC, 16. September 2015). Self-control is inscribed in female body figures (see Drummond and Rydstrøm 2004b) and was also described by the female students as a way emphatically to pay attention and cater to, before their own, the desires and needs of the other people with whom they eat. This is expressed in the assertion: 'When you eat, check the pots and pans' (group discussion students, HCMC, 16 September 2014). The female students expressed that they feel responsible for the convenience and comfort of others when sharing meals, explaining that they (the female students) would 'see the whole meal'. This means that they are supposed to be ready to anticipate the others' desires at any time. In formal meal situations, especially with guests, the female students feel required to follow the correct order of who to invite and serve first, relating this internalised meal practice to norms of seniority, gender and hospitality. Women are expected to know and act according to this script for sharing meals with other people.

The subjective accounts of the two older female informants document the historical development of the infrastructure of drinking and eating out, and how public social spaces for women have been facilitated in the context of economic reform and food system transformation since *đổi mới*. From a more global perspective, Julier asserts that '[r]estrictions on women's public eating dissipated as restaurants and eateries recognized their power as a consumer market' (Julier 2013: 342). Similarly, whereas strong Vietnamese ground coffee was exclusively reserved for men, the coffee industry, especially the instant coffee sector, discovered female consumers as a high-potential target group through the promotion of the sweet and less strong *cafê sữa đá* (interview, staff market research institute, HCMC, 21 September 2015). However, although these assessments suggest that the provision of female public space proceeded in a top-down manner due to capitalist market expansion, the subjective accounts of Anh and Ngoc referred to gendered socialising over food and drinks as an actual social practice conquering and acquiring newly emerging space in HCMC. The social norms of correct drinking patterns for women and men in public space are in the process of being renegotiated by the very social practices of women and men frequenting such places for leisure and recreational activities. In his study on Western-style fast food restaurants in Beijing, Yan shows that women were especially attracted to such venues because table manners were seen as more relaxed than the strict, gendered etiquette of male-dominated formal Chinese restaurants. In fast food restaurants women were explicitly expected to order their own food individually and to participate in conversations (Yan 2008: 92–93). Similarly, a comparison between the older and the younger generations' perspectives shows that public space for female entertainment and leisure has certainly widened. At the same time, the younger generation described how certain gender norms 'at the table' persist. This ongoing process of negotiation, which is constituted by the very practice of sharing food, echoes Bitter-Suermann's conclusion that '[w]omen in Vietnam are tasked with walking a line between modernity and traditional femininity, partly represented through what they eat [and drink]' (Bitter-Suermann 2014: 4).

Generational dimensions of commensality

In general, commensality refers to eating together as well as sharing food together at the same table. In Vietnam, the preparation, display, and consumption of food constitute highly sensory social practices that aim to stimulate sensual delights as well as foster commensality as a form of social bonding (Avieli 2012). A nutrition teacher sums this up when talking about Vietnamese food culture: 'The Vietnamese attach special importance to eating style, and … use all of their senses to enjoy the food, including sight, hearing, smell and taste' (interview, nutrition teacher, HCMC, 22 August 2014). As will be discussed below, eating out is the frame within which the social practices of commensality are dynamically changing, and food sharing and drinking out in public exhibit a strong generational dimension (see also Chapter 9 on generational issues and how this relates to food waste in Shanghai). The social practice of eating with peer groups, co-workers and friends as a way of spending leisure time together shapes a process in which commensality is starting to shift slightly away from family and direct kin groups. This again will be elaborated by contrasting the older informants' perceptions and practices of eating away from home with the accounts of the student generation.

To begin with, Anh, the informant referred to above, discussed the role of the family in commensality by contrasting past and present:

> In Vietnam many, many years ago, I mean before 1975, we used to live as a family. A big family – with grandparents, parents and children. Three generations in one family. So food was very important for everybody in the house – food for the elderly, food for adults and food for the kids. The cooking was done by the housewife. After 1975 most women got a job and went to work. At noon nowadays they stay in the office. So they do not have time to care for the children. Today we seldom have lunch together. We have lunch at school or lunch in the office. So we just have dinner together and sometimes we do not even have dinner together because after school, children go to evening classes. They also go to school during the weekend. … In the past, every young lady was educated in how to cook a lot of recipes and how to serve them. And nowadays most girls don't know how to cook. They go out [for food] or their mothers cook for them but they cannot cook a meal.
>
> (Interview, HCMC, 5 October 2015)

According to this statement, the changing labour market following the socialist reunion of the country in 1975[7] had an effect on the family as the main caregiving institution. Intended to break with Confucian patriarchal hierarchies, socialist egalitarian gender ideals promoted female labour and not only had an effect on female professional opportunities but also rearranged the everyday sphere of the family meal and the reproductive roles involved. The changing structural context of professional and educational opportunities after 1975 and the subsequent economic developments reconfigured the ways in which people share food together. Involved in this structurally driven development of eating out are the social repositioning of the female and the transfer of reproductive knowledge and everyday food skills from generation to generation (for the case of Cambodia see also Feuer 2015).

Besides the growing irrelevance of certain food-related knowledge and skills depicted, the micro-practices of food sharing related to gendered and seniority-based norms are also changing due to the increasing adoption of Western-style meal formats. Returning to the examples of fast food restaurants and upmarket coffee shops, the female students from the group discussion above described the eating behaviour of women their age as more relaxed and less moderate when eating out and socialising among peers. While eating out with friends in fast food restaurants they would not care so much about, for example, rules such as the correct sequence of whom to invite and serve first, something that was mentioned earlier on as part of a detailed script regarding norms of seniority, gender and hospitality when sharing food. They related this perceived norm relaxation to the aforementioned emic expression 'check[ing] the pots and pans'. The need felt by young women to 'see the whole meal' (group discussion students, HCMC, 16 September 2014) has become less relevant to commensality among equals because in such locations a more Western-style meal culture is followed in which individual plates are more common than the multitude of shared dishes characteristic of Vietnamese cuisine. Eating finger food instead of eating with chopsticks and ordering individual plates instead of serving and being served from shared dishes tend to favour the development of individuality in the context of eating 'together'. This eating style requires less of the empathy and consideration of others' gustatory needs indicated by 'checking the pots and pans'. The social obligation inherent in food sharing seems to be shifting increasingly from family-based commensality and female reproductive care towards another sort of obligation: the doing of middle-classness through the performance and embodiment of individuality and modernity.

The postulated Vietnamese ideal, initially referred to by the nutrition teacher, of enjoying food by 'eating with all one's senses' stands in sharp contrast to the social practice of eating out in fast food outlets. Visiting a fast food restaurant seems to be less about the gustatory act than it is the affective experience: 'modern' consumption of entertainment, youth culture belonging, 'doing middle-classness', and social status aspiration (see also Yan 2008: 91). The signifiers of 'togetherness', belonging and distinction become less centred on the mere act of eating and sharing food together. Instead, commercial and individualised food orders constitute a prerequisite for hanging out together for hours pursuing a perceived middle-class lifestyle and global consumer culture.

In this regard, the market researcher talked about middle class families and also presented her personal account:

> [W]e [middle-income families] have money and we just want to make our kid[s] happy and, you know, it is a good place for a family catch up, for a family hang-out. So [the] kids can play together and we have some time together and everyone is happy. ... From middle class up they can go there often, but even for the lower income groups, they know it is [something their kids want]. So if [a] kid achieves something, behaves nicely, it is a reward for them to go there. For example, after the semester, with a good

result they take the kid there to have some fun, have fast food, so everyone is just happy.

(Interview, HCMC, 21 September 2015)

Socialising over food and drinks becomes a minor matter and takes on different qualities. Anh, one of the older informants, observed similar attitudes among the young: '[They] like pizza, they don't like to eat rice nowadays. Because when they eat, they can look at this [mobile phone] and they can play games with one another or surf the internet. This is not socialising as in the past. This is socialising with the machine' (interview, Anh, HCMC, 5 October 2015). Data collected through participant observation in various locations suggest this is a common phenomenon: the younger generation, occupied in groups or individually with online chatting, checking social media and posting selfies taken in the coffee shops or fast food outlets, constantly communicate to the world that they are at this very moment part of global consumer culture.

Hospitality, social status and the habit of over-ordering

So far, this chapter has presented diverse examples of food sharing as a means of identity construction. In the rapid development context of post-*đổi mới* Vietnam, the (re-)positioning of oneself as consumer and as woman or man is, alongside the crafting of youth culture, intimately connected to everyday food practices marking distinction and belonging. Hospitality is also a feature of this, as exemplified by the emic remark about 'checking the pots and pans'. Avieli (2012) has documented the inevitable link between food sharing and hospitality as a common form of reciprocal obligation in Vietnam. Hospitality in the context of having meals together relates not only to quantity but also to sharing the best pieces of food. General observations in upmarket restaurants showed that customers tend to order many dishes when having food together in a bigger group. Participant observations in respective restaurant settings during lunch service were sometimes followed by an opportunity to talk to waiters or restaurant managers after the peak lunchtime rush. One such opportunity arose with the manager of a Thai restaurant who explained that, especially during office lunch breaks, small groups of two to three office mates usually order individual plates, whereas bigger groups of five or more tend to order a variety of dishes to share together. A similar opportunity emerged at an upmarket vegetarian restaurant. Drawing on his work experience of nearly twenty years in the hospitality sector (the hotel business in particular) in HCMC, the waiter, Mr Huu, observed that due to recent decades of economic growth, people nowadays go out for dinner, lunch and leisure much more often than compared with twenty years ago. Today, customers can afford to spend a whole night out in a restaurant enjoying drinks, food and having fun together. According to him, the young generation especially enjoys going out to socialise, and young females enjoying drinks and food in restaurants and bars are a common sight today (interview, HCMC, 13 October 2015). Prompted by my general observation that customers tend to order many dishes when having food together in a bigger group, he explained the following:

> You know, the Vietnamese, they like to order too much food when they organise a party ... They order everything, too many dishes, and the whole menu.
> (Interview, Huu, HCMC, 13 October 2015)

According to Huu's experience, this kind of food ordering practice mostly applies in the event of a 'party', for example a birthday party, and is also prominently practised in the context of a business meeting:

> If they invite people for [a] business lunch or dinner they try to order as much as possible. They do that because they want to show off and they want to treat their customers well [to encourage] good business in the future. So they don't care for money. All they want is the best and what is expensive [on] the menu.
> (Interview, Huu, HCMC, 13 October 2015)

The social obligation of a business dinner or lunch was similarly mentioned by Vietnamese friends and colleagues informally talking about their experiences (field notes, HCMC, 1 October 2015). Generosity in the context of business meals was experienced many times first-hand by the researcher herself. Besides the idea of caring for and treating others well, over-ordering can also be read as a demonstration of social status or aspiration:

> Vietnamese people like to show off when they have a party. They like to show their wealth and they order [for] more than the expected guests. This depends on the money you have but in general we always try to avoid running out of food.
> (Skype interview HCMC/Hanoi, 13 October 2015)

This quote is attributable to the founder of one of the first food bank initiatives in Hanoi, a woman and former student in her early twenties. As the project addresses the social implications of over-ordering – an excess of food on the one hand and, on the other, unequal access to food – it will be discussed in more detail below. At this point it can be noted that over-ordering as a food sharing practice arising from hospitality norms is not a new phenomenon. What is new, however, is the social meaning ascribed to this practice in the midst of change. Besides extending hospitality, ordering a lot of food of the best quality, highest price, and of the culinary 'non-ordinary' demonstrates a social status-enhancing quality. Not only providing 'enough' food but enjoying and, especially, sharing 'delicious' food – as the proverb cited at the outset suggests – comes into play again in the performance of middle-classness.

Social inequality and emerging 'counter-cultures'

This habit of over-ordering caught the attention of a student group in Hanoi, prompting them to initiate a food project called *Hà Nội Đủ* ('Hanoi Enough')[8] in the city in the summer of 2013. A narrative interview with the founder and an introduction to the project through the *Viet Nam News* article, 'Student-run food bank helps to feed poor people in the capital' (*Viet Nam News*, 10 September 2015)[9] communicate the rationale behind the idea. Through frequenting restaurants herself, the project founder realised that much food goes to waste, especially in venues where buffets are served, as well as due to the aforementioned practice of over-ordering. At the same time, wandering around Hanoi, she observed poor and homeless people who search for food in public bins. The project developed from this manifest mismatch between two parallel societal phenomena: on the one hand, the wasting of food and, on the other, food scarcity experienced by poorer groups. The group wanted 'to balance the gap between the [food] surplus and the ones who need it' and to address very practically the widening inequality gap in Vietnamese urban society (Skype interview HCMC/Hanoi, 13 October 2015).

The implementation of the project, however, still faces many challenges. For instance, the target group – 'the poor' – at first refused to take food that was thrown away or leftover by others:

> They don't want to be looked down on because of being poor. At first we fought hard to persuade them to use the food. Because they thought that we were thinking of them as poor, as weak. They felt ashamed of themselves.
>
> (Skype interview HCMC/Hanoi, 13 October 2015)

This hesitancy to accept such food donations intriguingly demonstrates the close connection between food and identity, as eating 'food waste' symbolically embodies and demonstrates poverty and inferiority, going against one's concept of the Self.

A search of social media and local newspapers for further food waste management projects revealed but a handful. Most commonly they take the form of informal practices such as the direct reuse of leftovers in piggeries,[10] the distribution providing an income source for poorer groups in society. The aforementioned waiter Mr Huu explained that restaurant leftovers are commonly eaten by the restaurant staff, whereas the hotel business in which he worked for many years followed a very strict policy of throwing everything away. More ambitious waste management projects that hold a clear vision of social equity and environmental sustainability are rather scarce, one being the aforementioned project *Hà Nội Đủ* and another the social media campaign *Ăn hết rồi* ('Eat Up Food')[11]:

Vietnam and other South-East Asian countries focus on economic development and they forget the environment or helping others because we [Vietnamese society] are moving forward in terms of economic development. Maybe this is why not too much … attention is paid to social services and all the food projects were led by students. And when their student lives end, the projects … end [as well].
(Skype interview, HCMC/ Hanoi, 13 October 2015)

This quote from the founder of Hà Nội Đủ indicates the reflexive capacity of consumers and a growing distaste for food waste among the younger population who want not only to be party to a perceived modern lifestyle and global consumer culture, but also to connect with the goals and discourses of global consumer criticism, for example environmental sustainability and growing social inequity within the food system. Food waste is thus gradually becoming a material subject of social and inter-generational negotiation regarding the 'obligations' attached to middle-classness, for example performing social status by over-ordering versus taking on societal responsibility to alleviate growing inequality in terms of food excess and food access. Food waste as a characteristic of the global food economy is thus becoming embedded in local social practices of 'doing middle-classness'.

Conclusion: practices and structures of food consumption and social inequality

This chapter has shown that the diversification of the food system and emerging consumerism in Vietnam have reframed broader societal negotiation processes that go beyond narrow understandings of consumerism among 'the' urban middle class. The examples of changing food sharing practices presented here highlight the manoeuvring of gendered social spaces and inter-generational norms of reproduction towards new forms of 'togetherness' bound by a need to perform middle-classness through the demonstration of 'modern' lifestyle knowledge and the performative skills connecting the individual with global consumer culture.

As it opens up dynamic micro-processes of identity construction, Vietnam's development and macro-level transformations are highly enabling. At the same time, however, the downside of growing inequality is also manifested in the food sphere. In a context of economic growth that is not equally distributed but has led to new forms of poverty including food scarcity, the food practices of eating out and over-ordering come to represent the growing inequality and exclusion of those not benefitting from the economic developments.

Backed by rising purchasing power and the growing social obligation of consumption (of entertainment, of 'modernity', of the 'foreign'), food-related practices of social distinction and boundary-drawing reproduce the structural context of the food system. In the long run, the production-driven food system creates not only new consumption needs but also growing externalities in terms of social inequality. The presented practices of sharing and wasting food show that food is increasingly becoming a social marker of middle-classness. At the same time, eating or recycling the food waste of others clearly symbolises the ingestion of poverty and an inferior identity. Environmental sustainability and food-related health considerations are but two of the further indications that urban food consumerism in Vietnam prefigures. The explorative foray into urban food projects and discourses featured above shows that not only are forms of 'modern' consumption globalising but also that certain aspects of global counter-cultures, social responsibility and resistance to consumerism are also beginning to be embedded locally.

This work was kindly supported by the Austrian Science Fund (FWF). Furthermore, my thanks go to Petra Dannecker, Christiane Voßemer and the editors of this book for their reviews and encouragement.

Notes

1 Free English translation from 'cơm no, áo ấm; cơm ngon, áo đẹp'.
2 The project 'A Body-Political Approach to the Study of Food: Vietnam and the global transformations' is funded by the Austrian Science Fund (FWF) for a three-year period that started in January 2015. For further information see: http://ie.univie.ac.at/en/research/projects/a-body-political-approach-to-the-study-of-food-vietnam-and-the-global-transformations/
3 For a more detailed introduction to female labour under late Vietnamese socialism see, for example, Bélanger and Oudin (2007).
4 BBC News (2014): McDonald's opens first outlet in Vietnam's Ho Chi Minh. http://www.bbc.com/news/world-asia-26101009 (accessed 25 January 2015).
5 All informants' names are anonymised throughout the text.
6 For a more detailed discussion of the role of women in Confucian thinking, see for example. Drummond and Rydstrøm (2004a) and Ngo Thi Ngan Binh (2004).
7 For a more detailed introduction to female labour under late Vietnamese socialism see, for example, Bélanger and Oudin (2007).
8 See also https://vi-vn.facebook.com/hanoidu/ (accessed 19 January 2016).
9 Viet Nam News (2015): Student-run food bank helps to feed poor people in the capital. http://vietnamnews.vn/society/275579/student-run-food-bank-helps-to-feed-poor-people-in-the-capital.html (accessed 14 December 2015).
10 In this model, food waste from public canteens, for example, is collected for free and brought to sub-urban piggeries. Offering a minimal but stable income, this job is often done by poorer groups in society. See for example: http://hanoimoi.com.vn/Tin-tuc/Nong-thon-moi/748911/thoat-ngheo-tu-tan-thu-thuc-an-thua-de-chan-nuoi (accessed 19 January 2016); http://m.cadn.com.vn/news/nuoi-lon-hieu-qua-tu-thuc-an-thua-loi-ca-doi-duong-8432-64 (accessed 19 January 2016).
11 See https://www.facebook.com/AnHetRoi (accessed 19 January 2016).

References

Agriculture and Agri-Food Canada. 2014. 'Food Service Profile Vietnam.'
Avieli, N. 2012. *Rice talks: Food and Community in a Vietnamese Town*. Bloomington: Indiana University Press.
Beardsworth, A. and T. Keil. 1997. *Sociology on the Menu. An Invitation to the Study of Food and Society*. London: Routledge.

Bélanger, D. and X. Oudin. 2007. 'For better or worse?: Working mothers in late Vietnamese socialism.' In *Working and Mothering in Asia: Images, Ideologies and Identities*, edited by T. Devasahayam and B. Yeoh, pp. 106–135. Copenhagen: NIAS Press.

Bélanger, D., L. Welch Drummond and V. Nguyen-Marshall. 2012. 'Introduction: Who are the urban middle class in Vietnam?" In *The Reinvention of Distinction. Modernity and the Middle Class in Urban Vietnam*, edited by V. Nguyen-Marshall, L. Welch Drummond and D. Bélanger, 1–17. Dordrecht, Heidelberg, London and New York: Springer.

Beresford, M. 2001. 'Vietnam: The transition from central planning'. In *The Political Economy of South-East Asia. Conflicts, Crises, and Change*, edited by G. Rodan, K. Hewison, and R. Robinson, pp. 206–232. Melbourne: Oxford University Press.

Bitter-Suermann, M. 2014. *Food, Modernity, and Identity in Ho Chi Minh City, Vietnam*. Halifax: Saint Mary's University.

Bourdieu, P. 1977. *Outline of a Theory of Practice*. Cambridge: Cambridge University Press.

Bourdieu, P. 1984. *Distinction. A Social Critique of the Judgement of Taste*. Cambridge: Harvard University Press.

Bui Van Hung. 2004. 'Rural diversification: An essential path to sustainable development for Vietnam.' In *Reaching for a Dream: Challenges of Sustainable Development in Vietnam*, edited by M. Beresford and A. Tran Ngoc. Copenhagen: NIAS.

Cadilhon, J.-J., P. Moustier, N. Poole, Phan Thi Giac Tam and A. Fearne. 2006. 'Traditional vs. modern food systems? Insights from vegetable supply chains to Ho Chi Minh City (Vietnam).' *Development Policy Review* 24(1): 31–49.

Coles, B. F. 2013. 'Ingesting places: The embodied geographies of coffee.' In *Why We Eat, How We Eat: Ethnographic Approaches to Eating*, edited by E. Abbots and A. Lavis. London: Ashgate.

Drummond, L. 2004. 'The modern "Vietnamese Woman": Socialization and women's magazines.' In *Gender Practices in Contemporary Vietnam*, edited by L. Drummond and H. Rydstrøm, pp. 158–178. Singapore: Singapore University Press.

Drummond, L. and H. Rydstrøm. 2004a. *Gender Practices in Contemporary Vietnam*. Singapore: Singapore University Press.

Drummond, L. and H. Rydstrøm. 2004b. 'Introduction.' In *Gender Practices in Contemporary Vietnam*, edited by L. Drummond and H. Rydstrøm, pp. 1–25. Singapore: Singapore University Press.

Earl, C. 2014. *Vietnam's New Middle Classes: Gender, Career, City, Gendering Asia*. Copenhagen: NIAS.

Feuer, H. 2015. 'Urban brokers of rural cuisine: Assembling national cuisine at Cambodian soup-pot restaurants.' *ASEAS – Austrian Journal of South-East Asian Studies* 8(1): 45–66.

Figuié, M. and N. Bricas. 2010. 'Purchasing food in modern Vietnam: When supermarkets affect the senses.' In *Asian Experiences in Every Day Life: Social Perspectives on the Senses*, edited by D. Kalekin-Fishman and K. Low, pp. 177–194. Burlington: Ashgate.

Figuié, M. and P. Moustier. 2009. 'Market appeal in an emerging economy: Supermarkets and poor consumers in Vietnam.' *Food Policy* 34: 210–217.

Food and Agriculture Organization. 1990. *Food Consumption and Nutritional Status in Viet Nam*. Nutrition Consultants' Reports Series No. 82. Rome: FAO.

Giddens, A. 1984. *The Constitution of Society: Outline of the Theory of Structuration.* Berkeley: University of California Press.

Hansen, A. 2015. 'Transport in transition: Doi Moi and the consumption of cars and motorbikes in Hanoi.' *Journal of Consumer Culture.* doi: 10.1177/1469540515602301.

Higgins, R. 2008. *Negotiating the Middle: Interactions of Class, Gender and Consumerism among the Middle Class in Ho Chi Minh City, Viet Nam.* Doctoral dissertation submitted to the Department of Anthropology, University of Arizona.

Julier, A. 2013. 'Meals: "Eating in" and "Eating out".' In *The Handbook of Food Research,* edited by A. Murcott, W. Belasco and P. Jackson, pp. 338–351. London, New Delhi, New York and Sydney: Bloomsbury.

King, V., P. Nguyen, and N. Minh. 2008. 'Professional middle class youth in post-reform Vietnam: Identity, continuity and change.' *Modern Asian Studies* 42(04), 783–813.

Mennell, S., A. Murcott, and A. Van Otterloo. 1992. *The Sociology of Food: Eating, Diet and Culture.* London: Sage.

Mergenthaler, M. 2008. *The Food System Transformation in Vietnam. Challenges for the Horticultural Sector Posed by Exports and by Changing Consumer Preferences.* Doctoral dissertation, University of Hohenheim.

Ngo Thi Ngan Binh. 2004. 'The Confucian four feminine virtues (tu duc): The old versus the new – ke thua versus phat huy.' In *Gender Practices in Contemporary Vietnam,* edited by L. Drummond and H. Rydstrøm, pp. 47–73. Singapore: Singapore University Press.

Nguyen-Marshall, V., L. Welch Drummond and D. Bélanger. 2012. *The Reinvention of Distinction. Modernity and the Middle Class in Urban Vietnam.* Dordrecht, Heidelberg, London, New York: Springer.

Peters, E. 2012. 'Cuisine and social status among urban Vietnamese, 1888–1926.' In *The Reinvention of Distinction. Modernity and the Middle Class in Urban Vietnam,* edited by V. Nguyen-Marshall, L. Welch Drummond and D. Bélanger, pp. 43–57. Dordrecht, Heidelberg, London, New York: Springer.

Schwenkel, C. and A. Leshkowich. 2012. 'Guest editors' introduction: How is neoliberalsm good to think Vietnam? How is Vietnam good to think neoliberalism?' *Positions: East Asia Cultures Critique* 20(2): 379–401.

Taylor, P. 2004. 'Introduction: Social inequality in a socialist state'. In *Social Inequality in Vietnam and the Challenges to Reform,* edited by P. Taylor, 1–40. Singapore: Institute of Southeast Asian Studies.

Warde, A. 2014. 'After taste: Culture, consumption and theories of practice.' *Journal of Consumer Culture* 14(3): 279–303.

Welch Drummond, L. 2012. 'Middle class landscapes in a transforming city: Hanoi in the 21st century.' In *The Reinvention of Distinction: Modernity and the Middle Class in Urban Vietnam,* edited by V. Nguyen-Marshall, L. Welch Drummond and D. Bélanger, pp. 79–94. Dordrecht, Heidelberg, London and New York: Springer.

Yan, Y. 2008. 'Of Hamburger and social space: Consuming McDonald's in Beijing.' In *Food and Culture: A Reader,* edited by C. Counihan and P. Van Esterik, pp. 500–522. New York: Routledge.

4 Eating in vertical neighborhoods
Food consumption practices in Metro Manila condominiums

Czarina Saloma and Erik Akpedonu

Condominiums have become a visible marker of a changing urban landscape and contemporary culture in Metro Manila, a mega city of about 11.7 million people and the capital region of the Philippines. A condominium is typically a high- or medium-rise building consisting of several floors and residential units in different configurations of studio and one-, two- or three-bedroom ensembles. In Metro Manila, these vertical neighborhoods are distinct in at least two ways. First, condominiums are usually built for the middle and upper classes: according to World Bank (2013) estimates, the demand for condominium units comes from socio-economic groups which have disposable monthly incomes of at least PhP 30,000 (USD 652) and which comprise about 20 per cent of households in Metro Manila. Other costs which restrict condominium living to these classes are the high cost of the parking lot rental or purchase, and the monthly association dues for building management and maintenance which, depending on the types of amenities and the overall state of the property, can amount to a significant financial burden (Ang 2009). Second, the condominium makes it possible for many people to live closer to the workplace and to leisure places. The attractiveness of condominiums, especially those with shopping malls at their podiums, is not least connected to the rise of the malls and the chronic traffic problem in the mega city. Ever since Crystal Arcade, the first mall in Metro Manila, opened in Binondo in 1932, and especially after World War II, Metro Manilans have enthusiastically taken to the malls. Being fully air-conditioned, malls offer a cool, clean, exciting and novel retail and entertainment environment, while the surrounding public parks, sidewalks, playgrounds and open spaces gradually fell victim to commercial development, infrastructure projects, neglect and squatters. With no viable alternatives left to go to, quite apart from being increasingly hemmed in by ever-worsening gridlocked traffic, building offices, malls and condominiums in unison provided an obvious and convenient solution for both problems.

The condominium phenomenon is by no way limited or unique to Metro Manila, but has become a hallmark of most fast-developing societies in the Middle East, the Indian subcontinent, China and Southeast Asia. Having been long established in Latin America (for example, Brazil and Argentina), it has recently spread even to sub-Saharan Africa, in countries such as Nigeria, Ghana and Angola. There, as in the Philippines, condominium living is primarily for the elite and the middle classes, in stark contrast to Europe and parts of North America, where, with exceptions, high-rise living is more commonly associated with socialized housing. However, few are perhaps as integrated with shopping malls and commercial spaces as the ones in Metro Manila. Yet, we know very little about the consumption cultures in these condominiums. By placing urban residents in a new material setting that could remake practices, condominiums raise important questions about how we are to understand changing food consumption practices, which encompass food acquisition, preparation, eating meals and waste management. They also prompt questions about how we are to understand the link between food consumption and the transition to sustainable pathways relating to, among others, energy consumption and waste generation. To find answers, the chapter opens with a framework and methodology for understanding food consumption in vertical neighborhoods. It then describes the material and social setting of these neighborhoods, followed by a discussion of the food consumption practices in these places. The chapter ends with an outline of the implications of these food consumption practices on the creation of a more viable and liveable mega city.

A social practice approach to food consumption in vertical neighborhoods

In this chapter, we explore the notion that new spaces and reconfigurational processes lead to expanding social patterns and practices (Knorr Cetina 1995). Practices refer to a 'routinized type of behaviour which consists of several elements, interconnected to one another' (Reckwitz 2002: 249). These include knowledge and understandings, bodily and mental activities, and things and their usage. With regard to the role of practical and general understandings within social practices, Schatzki (1996) highlights the 'teleoaffective' elements of practices and how ends, tasks and purposes coincide with beliefs, emotions and moods. For Schatzki (2001: 3), actions could only be understood within their specific practical contexts as 'the social is a field of embodied, materially interwoven practices centrally organized around shared practical understandings'. A practice approach therefore entails examining images, artefacts and competencies, in one definition (Shove and Pantzar 2005); or in another, a consideration for the knowledge distributed between people, things and culturally grounded structures (Wilhite 2005).

A particular focus on the material dimension in social practices draws attention to technologies and everyday objects (Shove et al. 2012), which provide frames for human activity (Berger and Luckmann 1966). Technologies and objects often quietly disappear instead of physically constraining or enabling; yet, while we do not 'see' them, they determine expectations and behaviors by setting the scene for normative practices (Miller 2005), such as consumption for personal expression and creation of a self-identity (Warde

1997) and sustainable consumption in terms of the use of material and energy resources in daily activities (Shove et al. 2012). Yet technologies and objects do not just impinge on society from outside of society: structure, or the social rules that comprise norms, procedures, conventions and resources including material settings and objects could both produce and be reproduced at the level of practice itself (Giddens 1984). The maintenance of practices over time depends on 'the successful inculcation of shared embodied know-how' (Schatzki 2001: 3) as well as on their continued performance (Schatzki 1996). Moreover, it also depends on practitioners' differing qualities and degrees of commitment to the practice (Warde 2005: 138).

Since no practice is 'hermetically sealed' from other practices, lessons can be learned from other adjacent and parallel practices, innovations can be borrowed and procedures copied (Warde 2005: 141). Thus, while practices involve routines (Reckwitz 2002) and taken for-granted realities (Berger and Luckmann 1966), 'crises of practice' (Reckwitz 2002) can bring about significant changes, such as new routines (Sahakian and Wilhite 2014). In the case of vertical neighborhoods, the competencies, cultural resources and social experiences of residents with regard to specific food consumption practices can be modified by the material settings of the condominium, and vice versa. Residents could completely adapt to new settings on one occasion, and retain old food consumption practices in another, for example. They could accept new ways of food consumption as provided by the new setting, but on other occasions, critique these new settings and find alternatives to consumerist, unhealthy or environmentally unfriendly practices. This chapter highlights the significance of space configurations and appliances, as the material dimension of food consumption in condominiums, and the dynamic relation between different elements of a practice.

Methodology

Data for this chapter came from a study on the changing food consumption practices and patterns among the middle classes in Bangalore, India and Metro Manila, Philippines (see Acknowledgements). This study examined household food consumption, where a household was understood to be a group of people sharing accommodations and the same food. Usually this group corresponds to a family, but non-family members such as domestic helpers were considered to be part of the household if they eat in the house. This study had qualitative and quantitative components: key informant interviews collected information on household food consumption practices, including consumption practices of household members outside the household, as well as data on food quantities that went into the household (through purchase and acquisition) and out (as food wastes). The quantitative data were used for the material flow analysis (for a complete discussion of the study's quantitative methodology, see Burger Chakraborty et al. 2016).

In Metro Manila, the study was conducted in three phases over a period of fourteen months, from August 2013 to September 2014. In the first phase, thirty households living in residences of all types, namely, single detached homes, row houses, and apartments in both gated and mixed neighborhoods, and condominiums were studied. About a quarter of this sample (eight households) lived in condominiums and appeared to eat out more often and were less aware about solid waste management than those living in other residential types. The second phase focused on thirty households living in condominiums. Whenever possible, the interviews were conducted in the homes of key informants, thereby providing opportunities for direct observation. In the third phase, a five-household case study was conducted. These households, representing single detached homes in both gated and mixed neighborhoods, a row house, and a condominium, were selected from the thirty households that participated in the first phase of the study. While the quantification of food acquisition in the first phase of the study relied on estimates provided by key informants, the household case studies asked a member of the household to measure household food inflows and outflows over a seven-day period and to put the data in a logbook.

The study selected research participants using theoretical sampling (Strauss and Corbin 1998). The sampling looked for criteria that were found to be repeatedly present or noticeably absent in the study's initial review of technical and non-technical literature and colloquial notions of the 'Filipino middle class'. The first criterion was income, wherein the annual middle-class household income corresponded to a broad range from PhP 246,109 (USD 5352) to PhP 2,000,072 (USD 48,837) (Virola 2009). From among those with such household incomes, the next selection criteria aimed to capture the diversity of the middle classes in terms of occupation of household head, household size and composition, housing type, migration history, city of residence and stays or travels abroad.

The use of tools from industrial ecology and from qualitative social science research enabled a richer understanding of the material dimension of food consumption practices. This was made possible by an iterative analysis of the various competencies of condominium residents and the meanings they attach to food consumption, on the one hand, and the patterns from the material flow analysis on the other. This cross-disciplinary and mixed methods approach allowed us to address the methodological tension between statistics and narratives, which are both needed in apprehending a social practice (see also Chapter 10 by Papargyropoulou and Chapter 12 by Favis and Estanislao, this volume).

Vertical neighborhoods in Metro Manila

The development of multi-storey apartments in Metro Manila dates back to the 1920s when they first appeared in Manila, the country's capital city. During the Spanish colonial era, concentrated mass housing primarily for the working class was in the form of *assesorias*, two-storey row housing in the suburbs surrounding Intramuros (the Walled City), typically built of brick on the ground and wood on the upper floor, with individual entrances on street level (de Viana 2001).

Following the US takeover of the Philippines in 1898, a massive investment and construction boom soon set in throughout the country, particularly in Manila. Of significance for the development of high-rise accommodation was the systematic development of the, until then, largely rural suburbs of Ermita and Malate, south of Intramuros, as the preferred residential districts of the new colonizers, bureaucrats, business people and professionals.

Capitalizing on the sweeping view and famed sunset of Manila Bay, American urban planner Daniel Burnham in 1905 developed a masterplan for Manila which proposed a wide tree-lined boulevard along the bay all the way south to the port of Cavite (Lico 2008). It was along this magnificent boulevard that Manila's first apartment buildings, typically five- to eight-storey concrete structures with comparatively few units on each floor, rose in the 1920s and 1930s. Art Deco masterpieces such as the Michelle, Angela and Rosario Apartments, the North and South Syquia Apartments, the Admiral and Bayview hotels and others formed Manila's new 'skyline' and attracted a wealthy clientele of locals and foreigners. The apartment boom continued after World War II and soon spread to newly developed subdivisions in the immediate environs of Manila, such as Makati. 1959 saw the raising of the then existing maximum height limitation of 30 meters in the city of Manila to 45 meters (which was raised again in the 1970s), allowing for the now fast spreading of high-rise office- and apartment buildings all over the city from the 1960s onwards. If previously building height rarely exceeded eight to nine floors, elevation of up to twenty floors became the norm. The 1990s further witnessed the arrival of very tall condominiums of over forty storeys, propelled by new and cost-efficient construction technologies and ever-increasing land prices in an ever-more crowded mega city. Today, Metro Manila is among the most densely populated urban conglomerates in the world.

To date, high-rise condominiums continue to be built in great numbers and with surprising speed all over the metropolitan area, and have spread not only to secondary urban centers such as Cebu City, but are now even making inroads in rural areas: only a few years ago, a number of massive condominiums were constructed in Tagaytay, a small resort town with hardly any mid-rises. Currently, huge multi-storey condominiums are being planned in Baclayon, a small historic town in Central Philippines whose tallest structure to date is the local church tower. While the latter two cases apparently aim to maximize the tourism and resort potential of their respective locations, the current condominium boom in Metro Manila is primarily driven by overseas Filipino workers (OFWs) (Sahakian 2014), as well as business process outsourcing (BPO) and information technology (IT) industries.

OFWs (Overseas Filipino Workers) are Filipino migrant workers, with about 2.3 million currently working worldwide (Philippine Statistics Authority 2014). The remittances they send home not only fuel a significant degree of retail and consumption in the Philippines, but are also invested in real estate, either to be lived in by relatives, to be self-occupied upon return to their homeland or to be rented out as an additional source of income (see, for example, Advincula-Lopez 2005; Aguilar 2009; Akpedonu and Saloma 2011). The number of Filipino families with OFWs is increasing, from less than 1 million in 2000 to 1.2 million families out of 17.4 million in 2006. Of this number, almost half (47 per cent) belong to the middle class (Virola 2009), thereby making the group a prime target market for condominiums. Given that it is comparatively difficult, risky and time-consuming to find and buy a second-hand home or to build a new one oneself, buying a house in a condominium unit from one of the many developers in the country is comparatively easy, convenient and safe. Moreover, innumerable sales agents operate in almost all popular malls, and marketing is well-organized and highly professionalized. Thus, hundreds of real estate projects are being offered in malls, newspapers, magazines, online and other media at any given time and are readily available, mostly through pre-selling. To further increase the acceptability of condominiums, some developers advertise the pleasures of condo-living and interior design options in purposely published high-gloss magazines, such as *Condo Living*.

Another important source of buyers are young urbanites in the booming BPO industry, or the outsourcing of business processes to a third-party service provider, which emerged in the early 2000s. Being generally young and comparatively well-paid, buying condominium units on credit or renting one is usually well within the reach of this new class. In addition, being most often still unmarried, and working late or even night shifts, the small size of most units is no deterrent for them. Moreover, most condominiums are conveniently located within the same central business districts which also house their workplaces. Young professionals repeatedly mention that time lost through daily commuting to and from work is a major factor for buying or renting a condominium unit.

The combination of housing–workplace–retail of typical mixed-use developments accounts for the predominant distribution pattern of condominiums in Metro Manila. Apart from the historic core of Manila, where condominiums originated about ninety years ago, and where they continue to be built capitalizing on the view of the bay, preferred locations are now Central Business Districts, such as the Ortigas Center, Ayala Center and Fort Bonifacio Global City. Transportation and transit is a major locational factor for condominiums, which are typically situated along major circumferential roads which connect the metropolitan area, such as EDSA (Epifanio de los Santos Avenue) and C5 (circumferential road #5), or along mass transportation arteries such as the LRT (Light Rail Transit) and MRT (Metro Rail Transit).

The attractiveness of condominiums not only lies in their strategic locations in relation to the rise of malls and chronic traffic problems. With ever more crowded streetscapes and perceived or real rising crime rates, heavily guarded and surveilled condominiums offer a safe refuge from the traffic, noise, pollution and dangers on the streets, secluding their inhabitants in a well-managed, controlled and sanitary environment remote from the realities on ground level (David 2004). Moreover, condominium living offers its members access to amenities which most of them would not be able to easily afford or access otherwise, such as swimming pools, fitness clubs and children's playgrounds. Maintenance of public areas is usually assured, unlike in the 'outside world', where public space is commonly neglected, overrun by squatters or simply does not exist. Greenery, albeit necessarily limited, is well-manicured, and written rules of proper conduct and behaviour among owners and tenants are in place in many developments, whereby problems can be readily addressed to the management team. Thus, the flight to the condominium is not least a reaction to the inability of the State to maintain a degree of order and safety that would make life outside condominiums and gated subdivisions more attractive and pleasant.

Designing food consumption in vertical neighborhoods

The condominium is a product of design, defined as the intentional shaping of matter, energy and processes to meet a perceived need or desire (van der Ryn and Cowan 1996 in Gross 2010). In Metro Manila, this means not only designing residential spaces but also sites of consumption, especially when considering that three of the top real estate developers in the country offer residential buildings and malls as their major products (Hoppler Editorial Board 2013). The typical design of a condominium complex features a residential building and a mall, or a residential building with at least a supermarket or convenience store at the podium. The typical mall would have high-end dining options on the upper floors, food court stalls and a supermarket at the basement level, and casual dining and fast food outlets on the lower floors. Casual dining outlets may not be on the same quality level of a fine dining place, but the menu is more varied and of higher quality than the typical take out or fast food joints. The basic market segment for casual dining is the middle class and young professionals (Cruz 2015), the same group who are likely to live in condominiums. Condominiums built by real estate developers that are not in the mall business would also still have, albeit limited, commercial spaces on the ground floor that are rented out to laundry shops, convenience stores, banks, restaurants and coffee or tea shops. Since many condominiums are built along major roads, their residents have easy access to restaurants and stores.

Compared to apartment houses from the 1920s and 1930s, the average floor area per unit in present-day condominiums has shrunk significantly. Typical condominium units in Metro Manila are likewise smaller compared to contemporary condominiums units elsewhere. In Bangalore, India, for example, a one-bedroom unit typically measures 57 sq. metres. In contrast, a typical one-bedroom unit in Metro Manila offers only 32 sq. metres of space, while a condominium studio usually measures from 15 to 20 sq. metres. With the exception of the luxury segment, units today are typically studios or one-to-two bedrooms apartments with areas mostly ranging from 20 to 50 sq. metres.

In an effort to maximize profit by using interchangeable and standardized designs and layouts, designers and developers of the built environment turn a place into a 'geography of nowhere', where a nondescript everyplace could be anyplace (Kunstler 1993). Most units follow a standard layout similar to hotel rooms, with en-suite bathrooms facing an internal corridor. Units are typically very narrow; thus, where a separate bedroom exists, it can usually only accommodate a single bed, while a standard double bed would fill almost the entire room. Such bedrooms are typically placed on the outer side with the window, while the living area inside does not receive any direct daylight, or only indirectly via the bedroom. Living rooms likewise suffer from the narrow layout and typically consist of a small sofa focusing on a flat-screen television set (a non-flat screen model would occupy too much of the precious floor space), wall-hung on the opposite side. Apparently, in most of these units the idea of the planners is that entertainment within the unit is not by socializing with guests over dinner or in a *sala* set (living room furniture), but solitary consumption of television or computer-related activities. Apartments likewise usually come with a very small kitchen unit for washing dishes and cooking, opposite to which can be found the bathroom typically containing a shower, toilet and basin in an extremely confined, but efficiently organized space. Efficiency in the use of space is the main characteristic of all these units.

As average floor area has shrunk, so has the ceiling height. Unlike pre-war apartments, which were naturally ventilated and thus required high ceilings, today's condominiums are almost without exception designed for airconditioning and thus require low ceilings for efficiency. Compared to pre-war apartments, glass and window surfaces have decreased in contemporary condominiums, at least in the low- and mid-priced market segment. At the same time these windows often cannot be opened or only at a small angle, thus preventing natural ventilation. Subsequently, almost all modern units depend on constant artificial ventilation, with the associated high costs for electricity in a country that boasts the highest power rates in all of Asia (Sahakian 2014). Due to the lack of space in most units, laundry has to be done in special areas either on roof decks or in basements, where laundry is dried either in special individual cages or electric dryers. Some condominiums also provide storage space in covered cages on the rooftop, as there is no storage space available within units. Condominium residents appear to have accepted and conformed to these limitations. A resident explained: 'We stay in a condo, so we have a fixed area for cooking, and the sink is beside it and then the microwave. That's the style because it's a condo, so it's a small space.'

In particular, a key informant said the space allotment in the kitchen influenced their choice of stove to buy: 'Because there's space for that [stove] so we were thinking if we buy range only, we need to have a table built or a cabinet. It will be more hassle so we might as well buy one with an oven too, so I can bake.'

Condominium residents thus have to choose among competing domestic technologies, and often the choice boils down to the optimization of living space. A research participant who lives with his wife and two young children in a 17 sq. metre condominium unit admitted they 'buy appliances depending on the available space in the condo. We measure. It is important for the appliance maker to know how much space is available in housing units. Two centimeters would matter. There is a need for smaller appliances.'

For a condominium resident who stayed with her family in a house in a subdivision before getting married, the limited storage space means that 'instead of buying in bulk which I used to do when I was with my family, my husband and I now buy weekly. Whatever we can finish in a week.'

Condominiums have house rules set by the board of the homeowners' association and implemented by the building administrators. Among the most common rules which affect food consumption practices are: no cooking on liquified petroleum gas, no pets and no parties in the units. Another key informant shared:

> You tend to be more dependent on the microwave. Heating things rather than cooking from scratch. Space constraints. There is little inspiration to cook because of the limited space. I like to spread out when I cook ... I might be inspired to cook if I have a bigger space. If you have a bigger place and if you have a conventional oven, you would cook ... We are no longer using the turbo broiler because it accumulates heat and will then heat up the unit.

Food preparation at home in vertical neighborhoods

The preparation of meals in Filipino middle- and upper-class homes is influenced by the preference for speed and convenience, and the presence of domestic helpers. The first can be traced back to the introduction of the 1909 Payne-Aldrich Tariff Act which made the country an instant market of American surplus goods such as breakfast cereals, milk, cocoa, cheese and soft drinks during the American colonial period (Salazar 2012). Eventually, the availability of surplus goods created what has been termed the 'Spam culture' (Fernandez 2000, cited in Salazar 2012), which favors processed over fresh food. Today, the Spam culture defines what meals are eaten at home, in particular breakfast, which typically consists of instant, canned, packaged and processed foods.

Likewise, being middle class in Metro Manila is associated with having stay-in domestic help. The presence of domestic helpers means that households have some form of help for preparing warm and freshly cooked meals for breakfast, lunch and dinner, and for cleaning up after meals (Sahakian et al. in preparation). Indeed, for many Filipinos a meal is almost always a warm meal of rice and what is termed 'viand', or a meat or vegetable dish, or both.

While the 'Spam culture' is very much alive and well, especially among singles who live in condominiums and who regularly eat instant noodles along with a variety of processed and canned foods, condominium living unexpectedly offers many of its residents access to 'freshness'. In most non-condominium households, food supplies are usually bought once a week, with small purchases made during the week. These households also have space for a bigger refrigerator and pantry or food storage areas. However, a key informant who moved to a condominium from a house in a subdivision to save on commuting time for himself, his wife and two school-aged children, mentioned that the well-stocked supermarket at the ground floor means that their domestic helper could buy vegetables on a daily basis, rather than stockpile them at home.

However, for some residents moving to a condominium means letting go of domestic helpers. A professor – who moved from a row-house in a quiet working-class neighborhood to a one-bedroom condominium unit across from the university – lost the space needed for in-house domestic help. Her helper now comes once or twice a week to cook and to clean, thus reducing her access to freshly cooked food on a regular basis. In exchange, she regains privacy, as she no longer has to share the space with a live-in helper. Newly married couples living in condominiums also tend not to have stay-in domestic help.

Those living in a condominium have to further reckon with other considerations when preparing meals at home. Concerns about the high cost of electricity drive many condominium residents, who rely on air-conditioning units to cool their homes and electric stoves to prepare their food, to spend more time in the air-conditioned mall – where cool air is free. Moreover, the small space combined with non-existent kitchen exhaust system also means that many households try to avoid cooking in their condominium units. Filipino food preparation predominantly involves the sautéing of garlic and onion, which generates a strong smell. With inefficient ventilation, the smell tends to linger. A key informant tried to adapt to this constraint by reducing the number of times in which she prepares Filipino food. Instead, she cooked more of the 'easy-to-prepare food', as she explained:

> because when you have to prepare like pasta you just mix everything together in a pan or like or you should fry the ground meat or pork with tomatoes, onions just like that … like let's say you'll bake cauliflower compare that to *adobong kangkong* (water spinach cooked in vinegar, soy sauce, and garlic mixture). Or sautéed vegetables, I don't even know how to cook that, whereas broccoli or carrots or whatever you just fry it in butter, that's it. So less of the traditional Pinoy (colloquial term for Filipino); we don't eat that at home because it's more difficult to prepare.

Moreover, due to the limited space and rules prohibiting the holding of parties in the units, many research participants reported not entertaining guests at home, as one of the changes brought about by living in a condominium. A retired professor who moved in to a condominium said: 'I don't have friends over. Yeah. I used to have friends over in my house.' Instead, she met with family and friends in the many restaurants near her condominium. When guests do come, they do not stay overnight. According to a newly married key informant, guests 'don't really come over and when they do, they just stay there for a day, they don't stay in the condo, but usually in the big house, the family house'.

Eating out in vertical neighborhoods

Households in Metro Manila, across all residential types, make a clear distinction between 'weekday' and 'weekend' activities when organizing their food consumption practices (Sahakian et al. in preparation). Generally, the weekday preparation of food pertains to simple everyday fare. In contrast, the weekend, specifically Sunday, is a time when more elaborate food is prepared or enjoyed. In particular, working female key informants who delegated the task of cooking during the week to household helpers, see the weekend, usually Sundays, as an occasion to prepare special meals themselves.

An analysis of the frequency of eating out among the households that participated in the study's first phase shows that households in condominiums eat out more frequently than non-condominium households: on average, the condominium household eats out five times a week, while the non-condominium household eats out four times a week. With many condominium households not having enough space for domestic helpers and food storage, and aiming to save on electricity costs, many are eschewing daily cooking. Couples and those living alone reasoned out that it is always cheaper to buy prepared food than to cook for one or two persons, due in part to the fact that the retail pricing system still makes buying in bulk more price-efficient for the buyer. Nevertheless, the weekday/weekend distinction is kept: on weekends, meals are usually taken in fancier restaurants and with the whole family, while eating out on weekdays is usually done out of pure convenience and in fast food or less fancy restaurants.

Condominium residents are at the center of a global and speciality food trend widely introduced by shopping malls. Global chain restaurants offer gourmet Chinese, Indian or Japanese cuisine, pancakes, sandwiches and pasta, among other diverse cuisines. Although Philippine regional cuisines are only starting to get a bigger share of the food service industry in comparison with global food chains, their emergence is supported by rural–urban migration, which remains the main driver of population growth in Metro Manila. Restaurants in malls, especially the casual dining outlets, are well patronized as food variety and quality become main considerations in deciding where to eat. A key informant who eats out about six times a week and who had lived a considerable time abroad, explained how she factors in variety and quality. According to her, choosing what to eat is 'not a conscious criterion. You like the food and it turns out to be imported, or a Filipino regional delicacy'. She continued: 'I was glad that I was back to my comfort food. The sooner you get the chance to have Filipino food, why would you buy the typical American food? But if Filipino [food] is not available … it is not that I am going to die.'

Yet, because eating at the condominium food court during weekdays was not tied to leisure and was mostly seen as a matter of convenience, some condominium residents see the weekend as a time for eating home-cooked meals. Many singles and young couples living in condominiums have parents in Metro Manila who usually live in bigger homes. Still abstaining from cooking themselves, they come home for laundry and for home-cooked meals on weekends. Moreover, food that has been cooked beforehand in their relatives' homes is taken back to the condominium, kept in the refrigerator, thawed, heated and eaten throughout the week. Thus, a married key informant and her husband who live in a mid-rise condominium unit visit her parents who live in a subdivision every week, and her husband's parents who live a bit farther away every other week. During these visits, they would take food supplies and home-cooked meals with them back to the condominium.

Towards sustainable consumption

The food consumption practices of condominium households have implications not only for energy consumption, but also for waste management. The waste produced in Metro Manila in 2005 totalled 6.7 million kg per day, or an average of 0.5 kg per day for every household (Department of Environment and Natural Resources 2005, cited in The Guidon 2012). Of this, food and kitchen waste accounts for 45 per cent of the total waste. Data from the household case studies show that the per capita amount of food delivered to a condominium household is three times and five times more than that delivered to a household in a single detached home and a townhouse, respectively. With the most food delivered, condominium households had the least cooked food at home, most packaging waste and most leftovers thrown away. Conversely, townhouses and single detached homes, which had more home cooked meals than those living in condominiums, had a greater amount of organic food waste such as vegetable and fruit peelings.

The country's Republic Act (RA) No. 9003 (Philippine Ecological Solid Waste Management Act of 2000) mandates that local government units reduce waste collection through the effective implementation of waste segregation at source. As with many other laws in the Philippines, the implementation of RA 9003 varies considerably from one city, municipality and *baranga*y (the smallest administrative unit in the country) to another, and from one condominium to another. In Metro Manila, garbage is usually collected by contract companies, which use dump trucks to collect it at least twice a week. The implementation of the RA9003 means that *nabubulok* (biodegradable or 'wet') and *di-nabubulok* (non-biodegradable or 'dry') wastes are collected separately and on different days of the week. Yet segregation seldom happens within the home. The formal garbage collection system is complemented by informal scavengers, who ensure that very little recyclable waste actually ends up in the dumpsite. A resident of a Makati condominium revealed how this works:

> In our condominium, there is only one garbage chute. At night I see scavengers segregating ... the condominium management is not strictly enforcing waste segregation at the household level because in the evening somebody segregates it for us.

A resident of another condominium described how the condominium manages waste:

> Mine is organic, we really produce little organic. And then you have the mixed wastes, like you know all these things, and then I have the papers, and then the bottles, which they sell. Our building has been selling every Saturday, there's a big truck that comes up at the parking lot. And they buy from all the condos.

Domestic helpers, usually the ones being sent by many well-to-do households to waste management trainings organized by the local government, play a central role in the management of household waste as they segregate the recyclables for additional income. According to one resident:

> *Yaya* (nanny) takes the empty bottles, plastic, and sells them. There are six to ten *yayas* in [the] condo who do this together and sell to the *mangangalakal* (scavenger) who has a pushcart. The *yayas* go around the condo, and what they collect, they first put in the roof deck.

A quarter of the Metro Manila households in the first phase of the study reported that they do not separate the organic waste. Other households throw it out separately, compost it, feed it to pet dogs and cats or give it to someone else, usually domestic helpers, to use. Giving the food to collectors for piggeries was a frequent answer (Burger Chakraborty 2014). However, with the 'no pets' rule, the security service that prevents food collectors from going to the units, and the absence of domestic helpers, a significant amount of leftover food is simply thrown away by condominium households.

One can also get an idea of the impact of condominium living on the environment in terms of energy consumption from the electricity used by home and kitchen appliances. The prohibition against using liquefied petroleum gas in the majority of condominiums means that in most households the electricity consumed by cooking appliances outweighs the usage for refrigeration. The main energy-eating appliances among those who participated in the first phase of the study are the electric cooker, electric kettle, regular oven, rice cooker, electric water dispenser and coffee machine. In the sample, the average food-related electricity consumption is 2.7 kWh per capita per week. As a comparison, the total electricity consumption of households in the Philippines in 2005 reported by Worldenergy.org is 4 kWh per capita per week (for details, see Burger Chakraborty 2014). The food-related electricity consumption of condominium households would be higher if electricity used to run the air-conditioning units at home during meals or in the malls and restaurants were to be considered. Already, an estimated 15 per cent of households in Metro Manila own air-conditioning units, and often several per household (Sahakian 2014). Moreover, the electricity used to cool the garbage in many condominiums is also to be factored in food-related electricity consumption. Garbage in condominiums is usually placed at a collection point, the size of which depends on the number of residents. Since garbage is not being collected everyday but rather about twice or thrice weekly, some condominiums provide air-conditioning for this room to slow down the decomposition of biodegradable waste. In one fourteen-storey building with twenty-six units on each floor, the garbage collection room measures about 20 sq. metres and is equipped with a one horse-power window-type aircon. In another condominium complex, with nine seven-storey buildings, this room measures around 60 sq. metres.

The impact of condominium living on the environment is not only on food-related consumption; tall condominiums can drive the energy cost of neighboring structures. As is to be expected in a mega city where land is at a premium, condominiums typically maximize the allowable space permitted by law, or by exemptions granted, such as the floor-area ratio and number of storeys. With heights and volumes ever increasing, high-rise towers are now packed closely together, with many units being shut off from sunlight and natural ventilation. At the same time, tall condominiums cast entire low-rise neighborhoods or adjacent condominiums into the darkness of their long shadows, negatively affecting natural light and ventilation. Light and heat reflecting off glass buildings can also heat up neighboring buildings (see Sahakian 2014). Consequently, the electricity consumption of these neighboring structures increases, either to light or to cool buildings.

Nevertheless, there are at least two areas where vertical neighborhoods appear to have reduced the food consumption-related impact on the environment: in the less frequent use of cars for the purchase of food, and in the better management of food stocks. In the first phase of the study, many key informants reported that they usually drive to the wet markets or supermarkets. Since they typically buy food supplies for the week, they do not like walking with heavy grocery bags. Thus, while many condominium households had the most food delivered and the most packaging waste, the location of supermarkets and restaurants at the podium or within walking distance could mean that less fuel is used in the transportation of food, from supermarket to home. A key informant who now lives in a centrally located condominium in Makati, a city within Metro Manila, also recalled that in her former homes she had to drive to eat out since most free delivery services require a minimum purchase of PhP 500 (USD 11) worth of food, an amount she considered excessive as she would normally be ordering only for herself, or for herself and her husband.

Moreover, not being able to stock food because of the lack of space for storing food and not having to do so because of the presence of a supermarket on the building's ground floor allow condominium households to manage their food stocks better and therefore to not only save electricity for deep freezers, but to also minimize the throwing away of spoiled uneaten food. Those who lived in condominiums away from their families particularly identified being alone and in full control of food decisions as the reason for lessening the quantity of food they prepare and consume – and thus reductions in food waste.

Concluding remarks

The two distinguishing features of vertical neighborhoods in Metro Manila – as homes for the middle classes and as sites that closely combine home, work and leisure – render them as potent spaces for redefining food consumption practices. Condominiums allow people to abandon practices such as cooking and having family meals at home while at the same time equipping them with new competencies and meanings, such as new ways of preparing food. In particular, condominium design reconfigures the home, hitherto the setting of the preparation of food and its enjoyment, by moving cooking and eating practices out of the residential unit – creating a norm whereby residents are no longer expected to cook and eat at home. This is made possible by limited spaces, inadequate provision for passive cooling and natural ventilation, expensive electricity and condominium rules for preparing and enjoying food. However, in these vertical neighborhoods, the home is not completely displaced: practices around material settings and everyday objects are both the medium and outcomes of practice itself. New routines emerge as residents embrace the food consumption options offered by the condominium at one point, or innovate on familiar ways of consuming, at another. The vertical neighborhood is therefore about the broadening, not solely the displacement, of food consumption practices. For some, the lack of proper ventilation simply means avoiding sauteing with garlic, and shifting to baking or non-garlic recipes. For others, choices among domestic objects and technologies are determined by space considerations, and guidelines on the use of electricity for food preparation. Yet, while the choice for stoves has been narrowed to electric stoves, some residents expanded this to electric stoves with ovens, or with ceramic or induction cooktops.

By emphasizing spaces and everyday objects, the chapter also shows that condominiums in Metro Manila support food consumption practices that have consequences for the environment in terms of waste generation and energy consumption, especially the use of electricity for food storage and preparation of meals, and fuel in the transport of food from supermarkets to home. In our study, condominium households had the most food delivered and consequently the most packaging waste compared to those living in single detached houses and townhouses. The food-related electricity consumption of condominium households could be higher if air-conditioning during the enjoyment of a meal, whether at home or at the mall, is to be considered. Nonetheless, the less frequent use of cars for the purchase of food and the better management of food stocks associated with condominium living could augur well for the creation of a more viable and liveable mega city.

Acknowledgements

This work was funded by the Swiss Network for International Studies, coordinated by Suren Erkman, Shalini Randeria, and Marlyne Sahakian, and supported by the Institute of Philippine Culture of Ateneo de Manila University.

We also thank our fellow researchers, especially Ma. Christine Camata, Laura Burger Chakraborty, Suren Erkman, Marie Abigail Favis, Lorraine Mangaser and Marlyne Sahakian, the members of the households who participated in the study, and Joanna Czarnecka-Pfaff and Gudrun Lachenmann. Czarina Saloma, in particular, acknowledges with gratitude, the Alexander von Humboldt Foundation for a research fellowship at Bielefeld University.

References

Advincula-Lopez, L. 2005. 'OFW remittances, community, social and personal services and the growth of social capital.' *Philippine Sociological Review* 53, 58–74.

Aguilar, F. 2009. *Maalwang Buhay: Family, Overseas Migration, and Cultures of Relatedness in Barangay Paraiso*. Quezon City: Ateneo de Manila University Press.

Akpedonu, E. and C. Saloma. 2011. *Casa Boholana: Vintage Houses of Bohol*. Quezon City: Ateneo de Manila University Press.

Ang, A. 2009. *Guide to Homeownership*. Mandaluyong: A1 Publishing.

Berger, P. and T. Luckmann. 1966. *The Social Construction of Reality: A Treatise in the Sociology of Knowledge*. London: Penguin Books

Burger Chakraborty, L. 2014. '(Un)sustainable food consumption dynamics in Bangalore and Manila.' Report on the MFA Part – Phase I (unpublished).

Burger Chakraborty, L., Sahakian, M., Rani, U., Shenoy, M. and Erkman, S. 2016. 'Urban food consumption in Metro Manila: Inter-disciplinary approaches towards apprehending practices, patterns and impacts.' *Journal of Industrial Ecology*, 20(3): 559–570.

Cruz, E. 2015. 'Rise of the middle class and casual dining.' *The Philippine Star*, 31 May.

David, R. 2004. *Nation, Self and Citizenship: An invitation to Philippine Sociology*. Pasig City: Anvil Publishing, Inc.

de Viana, L. 2001. *Three Centuries of Binondo Architecture, 1594–1898: A Socio-historical Perspective*. Manila: University of Santo Tomas Publishing House.

Giddens, A. 1984. *The Constitution of Society: Outline of the Theory of Structuration*. Berkeley and Los Angeles: University of California Press.

Gross, M. 2010. *Ignorance and Surprise: Science, Society and Ecological Design*. Cambridge, MA: MIT Press.

Guidon, The (Student publication of Ateneo de Manila University). 2012. 'Going green.' *The Guidon*. August–September.

Hoppler Editorial Board. 2013. *The Builders: 12 Top Real Estate Developers in the Philippines*. www.hoppler.com.ph/blog/featured-real-estate-developers/the-builders-12-top-real-estate-developers-in-the-philippines-part-1-of-3 (accessed 15 June 2015).

Knorr Cetina, K. 1995. 'Theoretical constructionism: On the nesting of knowledge structures into social structures.' Unpublished manuscript.

Kunstler, J. 1993. *The Geography of Nowhere: Decline of America's Man-made Landscape*. New York: Simon and Schuster.

Lico, G. 2008. *Arkitekturang Filipino: A History of Architercture and Urbanism in the Philippines*. Quezon City: University of the Philippines Press.

Miller, D. 2005. *Materiality*. Durham, NC: Duke University.

Philippine Statistics Authority. 2014. *2013 Survey on Overseas Filipinos*. http://census.gov.ph/tags/2013-survey-overseas-filipinos (accessed 25 November 2014).

Reckwitz, A. 2002. 'Toward a theory of social practices: A development in culturalist theorizing.' *European Journal of Social Theory* 5(2): 243–263.

Republic of the Philippines. 2000. Republic Act No. 9003. An Act Providing for an Ecological Solid Waste Management Program, Creating the Necessary Institutional Mechanisms and Incentives, Declaring Certain Acts Prohibited and Providing Penalties, Appropriating Funds Therefore, and for Other Purposes.

Sahakian, M. 2014. *Keeping Cool in Southeast Asia: Energy Use and Urban Air-conditioning.* New York and London: Palgrave.

Sahakian M. and H. Wilhite. 2014. 'Making practice theory practicable: Towards more sustainable forms of consumption.' *Journal of Consumer Culture* 14(1): 25–44.

Sahakian, M., C. Saloma, and S. Ganguly. In preparation. 'What makes for a tasty meal in Bangalore and Metro Manila? The role of collective conventions and communities of practice.' *Journal of Consumer Culture.*

Salazar, J. 2012. 'Eating out: Reconstituting the Philippines' public kitchens.' *Thesis Eleven* 112: 133

Schatzki, T. 1996. *Social Practices: A Wittgensteinian Approach to Human Activity and the Social.* Cambridge: Cambridge University Press.

Schatzki, T. 2001. Introduction: Practice Theory. In T. Shatzki, K. Cetina and E. Savigny (eds), *The Practice Turn in Contemporary Theory.* London and New York: Routledge.

Shove, E. and M. Pantzar. 2005. 'Consumers, producers and practices: Understanding the invention and reinvention of Nordic walking.' *Journal of Consumer Culture* 5(1): 43–64.

Shove, E., M. Pantzar and M. Watson. 2012. *The Dynamics of Social Practice: Everyday Life and How It Changes.* London: Sage Publications.

Strauss, A. and J. Corbin. 1998. *Basics of Qualitative Research: Techniques and Procedures for Developing Grounded Theory.* Thousand Oaks, CA: Sage Publications.

Virola, R. 2009. '*Pinoy* middle class before the crisis!' www.nscb.gov.ph/headlines/StatsSpeak/2009/060809_rav_middleclass.asp (accessed 22 June 2014).

Warde, A. 1997. *Consumption, Food and Taste.* London: Sage Publications.

Warde, A. 2005. 'Consumption and theories of practice.' *Journal of Consumer Culture* 5: 131–153.

Wilhite, H. 2012. 'Towards a better accounting of the roles of the body, things and habits in consumption.' In A. Warde and D. Southerton (eds), *The Habits of Consumption*, pp. 87–99. Helsinki: Helsinki Collegium for Advanced Studies 12.

World Bank. 2013. 'Philippine economic update: Accelerating reforms to meet the jobs challenge.' www.worldbank.org/content/dam/Worldbank/document/EAP/Philippines/Philippine_Economic_Update_May2013.pdf (accessed 22 June 2013).

5 From beef to bananas

Consumer preferences and local food flows in Honolulu, Hawai'i

Elizabeth Louis and Kyle Datta

Walk into the produce section of any supermarket in Hawai'i and the first things you notice are the obvious displays of locally grown foods. Previously in the purview of organic food stores and coops such as Whole Foods and Kokua Market, the desire to eat local has led to a burgeoning of marketing around local foods. Today, not just supermarkets, but restaurants and other food-preparing enterprises are offering whole menus based on local foods, or at least identifying local ingredients in their dishes.

There are no clear numbers on how much local food is produced or consumed in Hawai'i. Hawai'i imports 90 per cent of its food, and exports 80 per cent of its agricultural produce (Hollier 2014). But a closer look at these numbers shows that a high percentage of the state's fresh produce is grown in the islands. Some products like macadamia nuts, pineapples and coffee are grown in Hawai'i, but are exported; others such as lettuce, leafy greens, vegetables and fruit can meet the market demand, but the price may not be right. While there is a dearth of data, there is a definite trend of increasing demand for Hawai'i-grown foods, based on increases in local food production and retail sales. Based on dollar value estimates, the production for local food increased from 8.1 per cent in 2010 to 8 per cent in 2012, and 8.6 per cent in 2014 (Datta 2015).

Given the widespread support for local foods in Hawai'i, this chapter explores the meanings of local foods, examines how willingness to pay for local foods comes about, and explores the connections between local food consumption and the political economy of agriculture in the Hawaiian Islands. It highlights that local food consumption is influenced by the state's plantation past, the values and food preferences of its multiethnic populations, the socioeconomic standing of the state's residents, and the influence of global alternative food movements (AFNs). It suggests that Hawai'i residents' local food consumption practices share similarities with other Western countries, but are unique in how their multiethnic spatial identities influence consumption. As far as local food production goes, the state's challenges share more with other post-plantation island economies which are constrained by high energy costs, limited natural resources and the high costs of labour (MacLennan 2014).

This chapter is organized as follows: the first section will review literature on local food within the larger discourse of Alternative Food Networks (AFN). It then examines the local food movement in Hawai'i, tracing the role of the plantation economy and the diverse contemporary factors at play in influencing local food consumption practices. This is followed by a discussion of our research findings on supermarket shoppers' willingness to pay for local foods on the island of O'ahu, which includes Hawai'i's capital city Honolulu, and its suburbs. The fourth section will discuss our research findings through the lens of Hawai'i's spatialized food identities, suggesting that consumers' local food preferences reflect the state's multiethnic character and its plantation past. The last section will link local food consumption to the political economy of local agriculture in Hawai'i, discussing the opportunities and barriers to local food production as Hawai'i explores ways to advance a more sustainable food economy.

Localism and Alternative Food Networks (AFN)

Localization of food production symbolizes 'food in place' (DeLind 2006: 127) and is associated with viable rural livelihoods, environmental sustainability and social justice. Local food gives the perception of safety, authenticity, trust, transparency, confidence, cultural authenticity and sovereignty of food production (DuPuis and Goodman 2005; Louis 2012). Local foods are part of the large and increasingly complex ethical foodscape broadly referred to as Alternative Food Networks (AFNs) (Freidberg 2010). AFNs have emerged as a backlash against the negative environmental, health and economic impacts of globalized agriculture. The actors that comprise AFNs represent civil society, non-profits, international advocacy groups, retail chains, food coops, peasant movements and activists representing consumers and producers. The strategies they employ include practising sustainable food production, reducing food miles, increasing access to healthy foods for low-income communities, creating fair trade networks, supporting environmental and social justice initiatives, promoting food sovereignty, creating local markets for food, increasing healthy foods in school meals, advocating for equitable trade policies and supporting small farmers among other things (DuPuis and Goodman 2005; DeLind 2006; Costa and Besio 2011).

'Local' has trumped organic in the articulation of an alternative ideal to resist the universalizing nature of capitalism (Hart 2006) and the negative impacts of the globalization of food and agriculture (DeLind 2006). Local represents a 'second generation' in AFNs, as organic has become codified and commercialized (DeLind 2006). Organic production has largely become geared to meet minimum standards while replicating the long-distance, monocultural and large-scale agriculture that it was originally created to oppose (Guthman 2008; DeLind 2006).

Research suggests that motivations to buy local in the United States of America (USA) are driven by consumers interested in supporting the local economy, ensuring fair prices for farmers, maintaining local farmland and supporting small family farms (DuPuis and Goodman 2005; Seyfang 2006; DuPuis and Gillon 2009; Onozaka and McFadden 2011; Aprile et al. 2012). Because local food is understood to be fresher, less processed and more nutritious, it is considered to be healthier and is therefore also valued by those who are concerned with their personal health (DeLind 2006).

Critical work on local food networks in the USA highlights the importance of exploring the contexts and the multitude of values and meanings attached to local food. As is the case with organic foods, 'local' as an indicator of civic and environmental citizenship is being watered down as large retailers capitalize on the market value of selling local foods (DeLind 2006). Furthermore, because the market is the main mechanism of connecting local food producers to individual consumers, local food does not come cheap and is still largely available only to those with means, staying out of reach of underserved communities and the economically worse off groups in the USA (DeLind 2006). The market has individualized the local food experience, deflecting attention away from conversations on the structural impediments that prevent those without the necessary resources of time, information and money to access local foods (Allen et al. 2003; Dupuis and Goodman 2005; McCarthy 2006; Guthman 2008; DeLind and Bingen 2008; McIntyre and Rondeau 2011).

Research on willingness to pay for local foods consistently throws up the fact that local foods are mainly preferred by higher educated consumers and are only within the reach of those who have higher incomes. In Vermont, for example, a willingness to pay study found that consumers were prepared to pay more for local milk and apples. The section of consumers most likely to do this was younger consumers with higher household incomes (Wang and Sun 2003; Batte et al. 2007). In Europe, a study of urban consumers' preference for olive oil labelled with Geographical Indication (GI) found that consumers' preferences are impacted by their age, income and education (Krystallis and Ness 2005).

There is very little research done on 'local' food consumption in the Asian context. Most studies conflate organic, local and sustainable foods (Voon et al. 2011; Roitner-Schobesberger et al. 2008). However, important parallels exist between consumers in Western countries and Asian consumers. In Asian countries, the initial uptake of organic/alternative foods generally occurs among educated and wealthier consumers. As in the West, health concerns are the primary driver for choosing organic foods, but the choice to buy local is often linked with other values such as a desire for environmental sustainability. Similarly, the primary barrier for eating organic foods is price.

In a qualitative study done on the trade-offs consumers make between local and organic foods in Shanghai, China, health concerns were the most important and organic food was seen to be more important than locally grown food (Siriex et al. 2011). In this case, the differences between organic and local were quite ambiguous. However, those who privileged both health and environment supported locally grown organic food (Siriex et al. 2011). Just 'local' food was equated with unsafe foods that had pesticide residues. Price was seen as the main barrier for buying organic food. Overall, consumers were driven by much less altruistic concerns related to the environment and more by their own personal health (Siriex et al. 2011). In another study investigating what motivates early adopters of organic food in Guangzhou, China, beliefs about organic foods as safe, healthy, tasty and environmentally friendly were the main motivating factors (Thøgersen and Zhou 2012).

In Thailand, a study among consumers of organic products found that such products are perceived as safer, healthier and environmentally friendly. Fears of pesticide residues in fruits and vegetables are high in the country, which makes organic a safer choice. Here again, consumers are older, educated and have higher incomes than consumers of conventional foods (Roitner-Schobesberger et al. 2008). In Malaysia, consumers of organic foods were affluent, well educated, and well informed about the benefits of a healthy lifestyle and affordability was not seen as a concern. The study showed a relationship between a strong positive attitude toward organic foods and perceptions of affordability, which exerted a significant positive effect on willingness to pay and actual purchases (Voon et al. 2011).

A history of Hawaiian islands foodscapes

Local food production and consumption in Hawai'i share cultural politics with the broader alternative food networks or movements (AFNs), which have become a global phenomenon. The movement in Hawai'i has some unique characteristics, influenced by its plantation past, however the underlying concerns are similar to AFNs elsewhere in the world (Costa and Besio 2011).

The material and discursive practices around local food are intricately tied to Hawai'i's complex past, which intertwines the story of Native Hawaiians with settlers, missionaries, colonialism and plantation eco-industrial economies (Costa and Besio 2011; MacLennan 2014). Hawai'i, the fiftieth state of the United States, is a group of nine tropical volcanic islands formed by a hot spot under the shifting tectonic plates in the Pacific Ocean. Surrounded by thousands of miles of Pacific Ocean, the Hawaiian Islands lie about halfway between Japan and the United States. Till the late 1800s, the islands were an independent constitutional monarchy, which was forcibly overthrown in 1893 by American planters and businessmen with the backing of the US government. The US annexed the islands as a territory soon after and it became a state in 1959 (MacLennan 2014).

Hawai'i's unique place in the Pacific exposed it to a continuous influx of people and biota. From the 1000s to 1600s, Polynesians settled in the Islands bringing with them taro, pigs and fowl. These earlier immigrations altered the islands' landscapes significantly, even before the advent of plantation systems. From the 1500s, before the arrival of Europeans (pre-contact Hawai'i), Hawaiian agriculture developed sophisticated irrigated and dryland agricultural systems along with marine resources such as fishponds and fish hatcheries. Hawai'i's population expanded and stabilized with agricultural intensification leading to a centralization of the power hierarchy. Agricultural intensification brought significant changes to the landscape, especially in the lowland forests and coastal areas (MacLennan 2014).

Early non-Polynesian settlers were Europeans and missionaries who came to the Islands in the 1700s. This second wave of humans caused drastic reductions in Hawaiian populations with the introduction of diseases and use of Hawaiian labor for European trade (MacLennan 2014). The introduction of ungulate animals and diseases also took a considerable toll on Hawai'i's stable agricultural systems and spawned another round of ecological change (MacLennan 2014: 22). With the arrival of the Europeans, Hawai'i became integrated into commercial Pacific trade, supplying whaling ships with provisions such as vegetables and pigs. Land closer to the ports was diverted for trade-related agriculture and away from subsistence crops (MacLennan 2014: 9). Trade integrated Hawai'i into world commodity markets, serving as a major revenue source for the Islands' rulers. This enabled the monarchy to consolidate and stabilize its finances and to gain recognition as an independent nation (MacLennan 2014).

The processes of intensification and centralized control of agriculture along with the creation of private property were critical to the development of the modern plantation economy. Sugar cane and pineapple were the main plantation crops, with smaller amounts of coffee and macadamia nuts (MacLennan 2014). Before the plantation economy, Hawaiian agriculture was still devoted primarily to subsistence food production such as taro, fish, sweet potato, breadfruit, bananas, chickens and so on. Except for the towns and ports where foreigners and Hawaiians were involved in the cash economy, most Hawaiian's lived in settled villages and practised subsistence-focused agriculture (MacLennan 2014).

It was the creation of a plantation economy in the mid-1800s that had the most drastic and long-lasting impacts on the Islands' natural environment, and in its political and socio-cultural arenas. Land was taken over and dominated by sugar plantations and the corollary cultivation of rice in the terraced areas formerly dedicated to taro. Ranching and dairy operations developed alongside plantations. Animals were used for fieldwork and harvesting, to supply meat and dairy products for plantations and local markets, and hides for export. By the 1920s, the Islands' resources including soil, water and forests experienced major ecological changes at an unprecedented rate, made possible by large infusions of capital and the conversion of Hawaiian agriculture into an eco-industrial complex, not dissimilar to other regions growing commercial crops for the world markets such as Fiji (MacLennan 2014).

The application of Western science and technology along with the creation of natural resource policies in the interests of economic development were key to the creation of the industrial monocropped plantation agriculture in Hawai'i. The sugar industry propelled Hawai'i into global markets (MacLennan 2014: 34). Missionaries turned into sugar capitalists who became so wealthy that they were eventually able to challenge the authority of the Hawaiian monarchy.

Land control became concentrated largely by the "Big Five" former missionary families turned sugar and ranching capitalists (Kēhaulani et al. 2008).

The shift to industrial agriculture engendered a violent dispossession of Native Hawaiians from their lands (Costa and Besio 2011) and led to a drastic demographic change in the makeup of Hawai'i's population. The plantation system's labour needs combined with the population collapse of Native Hawaiians due to mortality and low fertility rates from introduced diseases in the mid-1900s led to several waves of immigration of skilled and unskilled labour from China, Japan, Okinawa, the Philippines, Korea, Azores (Portugal), Puerto Rico, Russia, Germany, US mainland and the Pacific Islands. Feeding the labour force with their preferred foods was tantamount to maintaining a smooth-running operation. As land was diverted to other uses or had become fallow for lack of ability to cultivate it, importing food became a strategy. To that end, rice, salted salmon and other foods became integrated creating an eclectic 'local' food culture tied to plantation economies (MacLennan 2014: 197–200). In Hawai'i, the term 'local' is a cultural identity that is linked to ethnic groups who migrated to Hawai'i to work in plantation agriculture. Today it includes Native Hawaiians, as well as Caucasians who have settled in the Islands for generations, and plantation-era migrants such as Japanese, Chinese, Filipino, Portuguese, Puerto Ricans and their descendants (Costa and Besio 2011).

Towards the mid-1900s, Hawai'i's plantation economy experienced a major shift as plantation crops, especially sugar, fell victim to cheaper production costs elsewhere leading to a slow decline in sugar, pineapple and coffee production. Land concentration in the hands of a few corporations has had lasting impacts on the pattern of development in the Islands – from tourism and resort development, to residential development and agricultural production. Large tracts of land formerly used for plantations and owned by private companies became prime land for tourism and resort development, increasing property rates and limiting options for the state's growing housing needs (MacLennan 2014). As Haunani-Kay Trask, director of the University of Hawai'i's Center for Hawaiian Studies succinctly puts it, 'instead of growing sugar, now they grow hotels' (Goldberg 1996: 1). Legacies of the plantation economy are also visible in the irrigation systems, the patterns of residential development, controversies surrounding resort development and the dominance of the military. Irrigation systems that were created to provide water to plantations and ranches are still in existence. Settlements of people were determined by the location of plantations and their processing facilities.

Contemporary local food trends in Hawai'i

The revival of local and traditional foods and pride in Hawai'i-grown foods has become a key strategy for those who support improved economic, environmental and human health in the Hawaiian Islands (Gupta 2014).

With the demise of the plantation economy, the state sought to develop a more diversified agriculture to revive Hawai'i's agricultural economy, as well as to defend and support local agriculture and livelihoods (Suryanata 2002). Support for diversified agriculture was widespread as many, including lawmakers, had strong cultural ties to the Islands' plantation past. Even as the agricultural economy dwindled in the face of global competition, large areas of land continued to be devoted to agriculture (Suryanata 2002).

Popular constructions of local food in Hawai'i can be traced to the creation of Hawai'i Regional Cuisine, a trend spearheaded by chefs in Hawaiian resorts, catering to the palettes of high-end tourists and Honolulu's wealthy residents (Costa and Besio 2011). Hawai'i Regional Cuisine was inspired by Alice Walters, who popularized the farm-to-table movement and the notion of 'California Cuisine' with her restaurant Chez Panisse in the 1970s (Suryanata 2002). While the idea was initially to simply provide diverse, innovative and flavourful foods imbued with the multiethnic character of Hawai'i, sustainability became central to the idea of Hawai'i Regional Cuisine (Costa and Besio 2011).

The authenticity of Hawai'i Regional Cuisine rested on its emphasis on a connection to the land – the *'aina* – and its rural and agricultural places. The idea of *geogastronomia* put forth by Caldwell (2006) and Costa and Besio (2011) illustrates the affective power of local foods in evoking a sense of place in Hawai'i. Connections between food and place rest on geographic imaginaries that draw on Hawai'i's unique landscapes and history. The experience of eating local is infused with nostalgia for Hawai'i's uniqueness, and emphasizes both eaters' sense of belonging and their commitment to sustainability. The concept of *'aina* refers not just to food coming from the land, but also Hawaiian indigenous culture and the life-sustaining ability of the land (Costa and Besio 2011). Therefore, eating local is an active performance of place-making (Costa and Besio 2011).

There are other understandings of 'local' that may not conform to the popular ideas of local as locally produced or sustainable (Costa and Besio 2011). These are foods typically associated with the 'locals' in Hawai'i which are a combination of locally grown subsistence foods such as taro, fish, breadfruit, bananas, chickens and sweet potato, as well as foods not associated with health and sustainability such as imported or processed foods like Spam (a popular food item made from canned ham and distinct to Hawai'i's local diets), macaroni salad and white rice. Locally grown subsistence foods make up a relatively large part of people's diets in certain areas among certain ethnic groups, especially the Native Hawaiians, Pacific Islanders and local communities, and are important to their lifestyles (Gupta 2014). A study in 2012 found that 40 per cent of Molokai families' foods were sourced from subsistence activities. In these areas, the death of the plantation economy, the dearth of other jobs, and the high costs of food when compared to the mainland has forced poor communities to depend more on subsistence and food assistance programs (Gupta 2014).

Other aspects of the local food movement speak to several interrelated developments and concerns in the Hawaiian state. The revival of native Hawaiian culture and the movement for sovereignty have made traditional foods like taro a symbol of authenticity, sustainability and self-sustenance. In tandem with cultural revival are concerns about the deteriorating health of native and Pacific Island peoples largely concentrated in poor and underserved communities. A drastic change in their lifestyles due to alienation from their lands and traditional ways of subsisting, combined with poverty, lower access to education and employment, and a higher dependence on cheap, imported and unhealthy foods have given rise to an epidemic of diabetes, obesity, heart disease and other lifestyle-related health conditions (Liu and Alameda 2011). Native Hawaiians have the highest incidence of diet-related diseases (Johnson et al. 2004) leading many community leaders to promote local and traditional food consumption as a way to address these health issues (Kukui Maunakea-Forth 2012, Wai'anae, personal communication).

In Wai'anae, one of Honolulu's poorest communities where native Hawaiians are concentrated for example, a public health initiative called Hawaiian Diet was launched in the 1990s to promote more nutritious diets based on traditional foods. More recent developments such as MA'O Farms' youth food sovereignty initiatives and Ka'ala farms' programs on the revival of traditional and local foods, while empowering and educating youth and the community, are examples of the converging of food sovereignty, cultural revival, economic development, localization and sustainability (Suryanata 2002).

All of the above issues dovetail with concerns around Hawai'i's isolated geographical position with its dependence on imports of oil, food and other basic necessities making it more vulnerable in the face of human-caused and natural disasters Some estimate that Hawai'i has only a week's supply of food, its dependence on imported food is close to 90 per cent. Therefore the move to more self-reliance in food production has broad support because of fears that 'the barge will stop' bringing foods (Gupta 2014).

Examining local consumption choices among Honolulu's food shoppers

This section discusses the results of a study conducted to assess consumer preferences for buying local foods in Hawai'i. The study was commissioned by Ulupono Initiative and undertaken by Omnitrak Group Inc., a market research firm. Ulupono Initiative is a program by eBay founder Pierre Omidiyar, which makes for-profit, not-for-profit, and social investments in companies and organizations that can create large-scale impacts in its three-part mission of more local food, more renewable energy, and less waste in the Hawaiian Islands (Ulupono.com n.d.). Ulupono Initiative undertook the study to better understand the consumer side of supply-and-demand for local food across Hawai'i's food system. The study was designed to help Ulupono Initiative develop investment strategies to further promote local food production and consumption in the state.

The research, conducted in 2009, investigated consumers' willingness to pay for local staple foods signified by six commodities – beef, eggs, milk, lettuce, tomatoes and bananas. The study specifically sought to assess the attributes of local products that influenced consumers' purchasing decisions. Methods included focus groups, surveys and open-ended questions.

Focus groups were used to determine which commodities best represent the local food market and faced competition from imports. Milk, eggs, beef, lettuce, tomatoes and bananas were chosen because they represent high-volume purchases made weekly by consumers across all demographic groups in Honolulu. These products have both local and imported options in grocery stores.

The surveys asked consumers to grade and prioritize the attributes that would influence their decisions to purchase local food. The attributes presented in the surveys were price, place of origin, growing/production conditions, length of time from farm to shelf, texture, availability, taste, appearance, whether natural (grown without hormones or organically), color and variety. Since 'local' has many nuanced meanings, the study used the attribute 'Place of Origin' to determine the relative weight that consumers place on knowing that their food was produced in Hawai'i.

Honolulu, being the largest city in the state of Hawai'I, served as a good case study to understand consumers' perspectives on local food. A sample of 1196 adults who were the primary shoppers for food in a household was selected for interviews. Consumers were from five areas of O'ahu: Greater Honolulu, Central O'ahu /North Shore, Waianae, Windward side and Leeward. Each participant represented a unique household. Participants were randomly selected and reflected the underlying demographic groups in Honolulu. The 2010 census for Honolulu County lists Asians as being the largest population group at 43.9 per cent, followed by Caucasian at 20.8 per cent, Native Hawaiian or Pacific Islander at 9.5 per cent and individuals identifying as two or more races at 22 per cent (US Census Bureau 2010). Among the Asians, Japanese make up 15.7 per cent, Filipino 14.9 per cent and Chinese 5.4 per cent (US Census Bureau 2010). The sample of shoppers for the study was 8 per cent Hawaiian, 9 per cent Filipino, 27 per cent Caucasian, 22 per cent Japanese and 34 per cent other groups.

A conjoint analysis was used to determine the relative importance of each attribute for each commodity. The analysis was used to develop a consumer preference choice model. Results show that local food has broad community support: 74 per cent of consumers believe it is extremely important for Hawai'i to grow its own local food. Native Hawaiians have the strongest desire for local food at 83 per cent. *However, there is large gap between consumer desire for local food, and the ability of those consumers to afford or have access to affordable local food.*

The analysis of consumers' willingness-to-pay responses highlight that while the values underlying local food preferences are diverse, actual consumption is defined primarily by the affordability of products, and secondarily by factors such as availability of foods and knowledge of local food options, and cultural preferences for certain foods. This ties into arguments made by critical food studies theories that political-economic factors largely govern people's choices to participate in the local food economy. While consumers are increasingly seeking out local foods, local foods are still largely beyond the reach of poor and underserved communities, unless these foods are accessed outside the market as in foraging and fishing.

The consumer preference model found that price parity, access and the ability to identify (know) that the product was local, were the three most important attributes that defined customer choice across all commodities. While price parity was important in all commodities, it was most important for high-volume staples like milk, beef and eggs. Place of origin, whether the commodity was local or not, was the next most important attribute for eggs, bananas, lettuce and tomatoes. For beef, whether the product was 'natural' (for example, hormone free, non-feedlot) was the second most important attribute. Local bananas and tomatoes have high market share (65 per cent and 80 per cent respectively), while local milk and beef have less than 10 per cent market share. The relative price competitiveness of local production in relation to imports seems to impact these market shares. However, consumers are also loyal to local products. For example, apple bananas, a local small banana variety originally from the Philippines, are very popular among Oʻahu's consumers, despite the higher prices.

Price and availability were consistently the top two reasons that consumers in all demographic categories did *not* buy local food. Staples like milk, eggs and beef were bought at supermarkets or big box stores like Costco, a members-only wholesale enterprise that sells a wide range of products and fresh produce in large quantities at relatively lower prices. However, local brands are not always available and the limited availability of local food in these stores is a concern for consumers. For fruits and vegetables, availability was less of a concern in part because 16–24 per cent of consumers purchased these commodities at farmers markets rather than through traditional retailing channels.

The dominant demographic driver for purchasing local food is education and income, which are highly correlated in Honolulu. The urban core has the highest education and income levels, with a higher population of Caucasian and Japanese. The suburb of Waianae, on the other hand, is a historically underserved suburb with a high concentration of Native Hawaiians. Its residents have the lowest income and education on Oʻahu. Consumers with at least a partial college experience or a college education/graduate degree are more likely to purchase local food across most, but not all, commodities. Female middle-class shoppers earning between USD 50,000–100,000 per year are the most likely to purchase local food.

While Oʻahu consumers are price-conscious, the study's findings indicate that shoppers are willing to pay more for local products, but often have trouble distinguishing between locally grown and imported products. Consumers said that when it comes to local foods, they value quality especially in perceived freshness and taste; they trust local farmers to grow food that is safer to eat; and they want meat that is free from hormones and antibiotics. When products with the characteristics noted above are available and easy to identify in-store, the results indicate that some shoppers would be willing to pay up to USD 1.25 more per dozen of local eggs, up to USD 1.75 more per pound of local apple bananas, and up to USD 2.13 more per pound of local rib-eye steak than the brands imported from the mainland United States.

Spatialized food identities

The findings suggest that motivations to buy, grow or eat local foods are closely linked with the historical spatialized food identities in Hawai'i. The cultural preference for foods illustrates that the idea of *terroir* is relevant to understand consumer choices in the Hawaiian context. Products specifically grown in Hawai'i take on a value beyond the economic or environmental benefits, to become a social good, imbued with a sense of pride and ownership (Costa and Besio 2011).

An analysis of data across the five ethnic groupings in the study – Hawaiians, Caucasians, Filipino, Japanese and others – found that across all six commodities, Hawaiians consistently placed greater importance on the origin of their food compared to all other ethnic groups. Caucasians, many of whom were not born in Hawai'i, placed the lowest importance on having their food originate in Hawai'i. This suggests that discourses of pride, sovereignty and connection to the land among Native Hawaiians influence their attitude toward food grown in Hawai'i, more than any other cultural group in the state.

Local beef consumption is another example of the effect of spatialized food identities on buying choices of consumers. Less than half of the survey participants purchased local beef; of those that did, only 25 per cent bought local beef the majority of the time. These high-frequency buyers differed from the underlying demographic mix of the total sample. Lower income (<USD 25,000 per year) Hawaiian and Filipino ethnicities disproportionately purchased local beef. Hawaiian and Filipino represent 8 per cent and 9 per cent of the total sample yet there were 17 per cent and 15 per cent respectively of the high-frequency local beef consumers. The Waianae region represented only 12 per cent of the total sample, but 20 per cent of the frequent local beef buyers. Hawaiian, Japanese and Filipino had the highest predicted market shares (46 to 50 per cent), if price, quality and consistency were addressed. The over-representation of Hawaiians and Filipinos in preference for local beef speak to their strong connections to ranching and plantation *paniolo* cultures on the Big Island of Hawai'i. Conversely, more highly educated and wealthy Caucasian ethnicities were 27 per cent of the sample, yet only 13 per cent of the frequent beef buyers.

Apple bananas, the local banana variety, are a distinctly Hawaiian fruit and available in farmers markets, grocery stores and food stands all over the state. They were extremely popular and nearly 86 per cent of the participants purchase local bananas. Hawaiian, Japanese and Chinese all exhibit very strong market preference (greater than 73 per cent predicted market share). The

Filipino ethnicity was disproportionately likely to be infrequent buyers of local bananas (less than 25 per cent of the bananas they purchase). Filipinos were 9 per cent of the total sample but 13 per cent of the infrequent buyers, and had the lowest predicted market share. The reason for this needs further exploration.

Conclusion

This chapter concludes by examining the challenges linked to the production of local foods. The study on consumers' willingness to pay highlighted the importance of making local food competitive with imports, as price was the main reason that consumers were not buying local food. In the study, when relative price parity was established (for example, within 10 per cent) and consumers were aware of the origin of the products, then the predicted market share of local food rose to 55 to 70 per cent for bananas, tomatoes, lettuce and eggs. For beef and milk, the predicted market shares rose substantially from their current market share (to 44 per cent and 35 per cent respectively), but considerably lower than the others, highlighting how important price is for these staples.

To compete with cheaper imports, small farms would have to achieve economies of scale by lessening production costs through labour-displacing technologies and addressing other barriers. There are several challenges on this front, discussed below.

Agricultural production begins with the foundational inputs: land rights, healthy soil, water rights and labor. In each case, there are major concerns facing Hawai'i. Land ownership is highly concentrated and agricultural lands are under considerable pressure from real estate development (MacLennan 2014). Urban encroachment on agricultural lands, and the expectation that buffer zones will be created between agricultural activities and settlements, is further limiting agricultural development in highly populated districts in O'ahu, Maui and Kauai (Koberstein 2014). As a result, farmers both large and small have difficulty accessing affordable farmland with sufficient tenure to make major improvements.

Farm labour in Hawai'i is typically insufficient for the needs of the existing agricultural base, resulting in extensive reliance on immigrants from Asia, Micronesia and Central America (Mostafanezhad et al. 2015). The recent scandal of Aloun farms, a well-respected local family farming enterprise, which was charged with human trafficking of Thai guest workers, is testament to the major constraints faced in finding adequate labour in the state (Hawai'i Island Journal 2011). Unlike the mainland United States where seasonal migrants from Latin America fulfill much of the labour needs, Hawai'i's geographical isolation and high cost of living make it a challenge to attract affordable labour. On the Big Island, for example, many small farming operations depend on unpaid labor such as 'woofers', to fulfill their labour needs cheaply. Woofers are those who labour on organic farms within the World Organization of Organic Farms (WOOF) Network. WOOF offers young people opportunities to travel and gain agricultural skills with food and board, in exchange for labour. Often these 'gentleman' farms are subsidized through lucrative non-agricultural careers and supplemental income from retirement (Mostafanezhad et al. 2015). Even if farmers are not wealthy, farming cannot be relied on for subsistence, but more as a means to create additional income.

Water rights and access, both key for sustenance in an agricultural economy, are also part of contentious issues tied to the legacies of the plantation system. Sugar was a thirsty crop and historically all of the Islands' water resources were channeled through canals and ditches to the plantations (MacLennan 2014). Climate change has reached Hawai'i, and the long-term trend has demonstrated a consistent reduction in available rainfall, which is likely to accelerate over the course of the century (Leong et al. 2014). Furthermore, soil fertility in the former sugar and pineapple plantation lands is extremely low due to over a century of intensive agricultural practices. Pollution from herbicides and soil erosion bear the mark of Hawai'i's eco-industrial heritage (MacLennan 2014).

While land and labor are often talked about as making Hawai'i agriculture non-competitive, the cost of energy has received less attention, but is a crucial input into a farm, and one that adds to the cost of production (Hollier 2014). As Richard Ha, a successful tomato farmer in Hamakua Springs on the Big Island, notes: 'In the final analysis it's all about costs. The customers will go to where it's cheapest. So if our electricity was cheaper, the people would buy the stuff made here' (quoted in Hollier 2014).

There are other factors beyond land, labour, water and energy that are important to promoting local food production and consumption. For example, the supply chain across nearly all major commodities is either broken or under severe stress. The supply chain for cold storage, processing and distribution needs to be modernized to meet the US legal requirements for food safety and to efficiently move produce from farm to retail outlet. One example is the local beef industry which has faltered over the years, suffering from inconsistent quality and supply. The ubiquitous availability of low-cost, consistent-quality grain feed beef has relegated local beef to either high-end niche markets or ethnic markets.

However, investment in the supply chain is rising. For example, in meat processing, public–private partnerships have provided the capital to modernize processors in O'ahu and Hawai'i counties. The Hawai'i County processor was able to pass the national standards for food certification, which has opened up the markets in national chains such as Costco, Walmart, Whole Foods and Safeway to sell local meat. In March 2014, Parker Ranch along with Ulupono Initiative, launched the Paniolo Cattle Company, a joint venture to develop a state-wide grass-fed beef industry to tap into a growing demand for high-quality and affordable grass-fed beef (Parker Ranch 2014).

How to increase local food production is a subject of much debate and discussion. This question necessarily concerns how and where food is produced, whether the aim is total self-sufficiency of food production, or whether the focus should be on some fresh foods and a few value-added products. In addition to small farmers, there are several stakeholders interested in supporting the local movement including chefs, social justice activists, individual and family farming enterprises like Nalo Farms, ranching operations like Ponoholo Ranch raising grass-fed beef, to non-profits like MA'O Farms and the Kohala Center, investors like Ulupono Initiative and the trusts set up by Kamameha Schools, and Parker Ranch, two of Hawai'i's largest landowners. In addition several state, local and research organizations such as the Department of Agriculture, University of Hawai'i's College of Tropical Agriculture (UH CATHR), Hawai'i Agribusiness Development Corporation, High Technology Development Corporation, the US Army and the Office of Hawaiian Affairs are all stakeholders in policies and programs that advance local production and consumption.

Given all the issues discussed above, there is a growing consensus that aiming for total food self-sufficiency is not realistic given consumers' preference for globalized foods and the limitations of Hawai'i's tropical agriculture. Rather, the focus is now on replacing the importation of fresh produce and proteins with locally grown foods. As Hollier (2014) notes:

> If all of these elements – energy, technology, coordination and more – come together, they probably won't be enough to make Hawai'i completely self-sufficient in food – something we haven't seen since the days of the Hawaiian kingdom. But that doesn't mean we can't rejigger the system to make local farming more productive and enable it to grab a bigger share of locally eaten fruit, vegetables and other foods.

However, there are ideological differences as to what needs to be done to increase local food production. Jason Van Tassell from Parker Ranch, one of the largest landholders in Hawai'i and a stakeholder in Hawai'i's diversified agriculture, states:

> When the general public thinks of sustainable farming, I think they see a guy out in a field hand-picking vegetables and hoeing weeds. But I think sustainability means using every bit of technology that's out there, putting it to use to improve productivity.
>
> (Quoted in Hollier 2014)

Others are more tied to ideas of small-scale agriculture and the promotion of community values. Jon Osorio, Professor of Hawaiian Studies at the University of Hawai'i at Mānoa and a well-known activist and cultural practitioner, shares different views:

> I do not believe that we cannot cultivate food profitably in Hawai'i, though it might not be profitable enough for some. And I do not believe that economic activities can only work when large companies and major investment funds are privileged. I think that is a myth spawned by the Big

Five (dominant plantation-era families) back in the late days of the Kingdom, and I think we should hurl that myth into the pile of bagasse where it belongs.

(Hawai'i Book and Music Festival 2011: 5)

Despite these differences in opinion and the formidable obstacles to local food production discussed above, Hawai'i's local food sector is beginning to revitalize after nearly two decades of decline. Landowners, particularly the larger trusts and family-run ranches, have committed land and water to local food production. Numerous state policies support local food labelling, identification and awareness. Education is becoming an important aspect in promoting the local food movement connecting local schools to farms and businesses in the state while increasing awareness of the benefits of eating healthier and less processed foods (Nicole Milne, personal communication, Hamakua 2012). Of note are the school garden programmes in the state that are educating local children about the benefits of growing and eating fresh local foods. Within the decade, Hawai'i could see a resurgence of local food production to meet the consumer demand that is so clearly evident.

References

Allen, P., M. FitzSimmons, M. Goodman and K. Warner. 2003. 'Shifting plates in the agrifood landscape: The tectonics of alternative agrifood initiatives in California.' *Journal of Rural Studies* 19(1): 61–75.

Aprile, M., V. Caputo and R. Nayga. 2012. 'Consumers' valuation of food quality labels: The case of the European geographic indication and organic farming labels.' *International Journal of Consumer Studies* 36(2): 158–165.

Batte, M., N. Hooker, T. Haab, and J. Beaverson. 2007. 'Putting their money where their mouths are: Consumer willingness to pay for multi-ingredient, processed organic food products.' *Food Policy* 32(2): 145–159.

Caldwell, M. 2006. 'Tasting the worlds of yesterday and today: Culinary tourism and nostalgia foods in post-Soviet Russia.' In R. Wilk, *Fast Food/Slow Food: The Cultural Economy of the Global Food System*, pp. 97–112. Lanham, MD: Altamira Press.

Costa, L. and K. Besio. 2011. "Eating Hawai'i: Local foods and place-making in Hawai'i Regional Cuisine." *Social and Cultural Geography* 12(8): 839–854.

Datta, K. 2015. 'Feed ourselves.' Paper presented at the HCPO/HIGICC Island Futures Conference, Honolulu, October 2015 with Ulupono Initiative. Available at: http://www.islandfutureshawaii.org

DeLind, L. 2006. 'Of bodies, place, and culture: Re-situating local food.' *Journal of Agricultural and Environmental Ethics* 19(2): 121–146.

DeLind, L. and J. Bingen. 2008. 'Place and civic culture: Re-thinking the context for local agriculture.' *Journal of Agricultural and Environmental Ethics* 21(2): 127–151.

DuPuis, E. and D. Gillon. 2009. 'Alternative modes of governance: Organic as civic engagement.' *Agriculture and Human Values* 26(1–2): 43–56.

DuPuis, E. and D. Goodman. 2005. 'Should we go "home" to eat? Toward a reflexive politics of localism.' *Journal of Rural Studies* 21(3): 359–371.

Freidberg, S. 2010. Commentary. *Environment and Planning A* 42(8): 1868–1874.

Goldberg, C. 1996. 'As sugar fades, Hawai'i seeks a new cash crop.' *The New York Times*, 9 August 1996. www.nytimes.com/1996/08/09/us/as-sugar-fades-Hawai'i-seeks-a-new-cash-crop.html?pagewanted=all (accessed 15 June 2015).

Gupta, C. 2014. 'Sustainability, self-reliance and aloha aina: The case of Molokai, Hawai'i.' *International Journal of Sustainable Development and World Ecology* 21(5): 389–397.

Guthman, J. 2008. 'Neoliberalism and the making of food politics in California.' *Geoforum* 39(3): 1171–1183.

Hart, G. 2006. 'Denaturalizing dispossession: Critical ethnography in the age of resurgent imperialism.' *Antipode* 38(5): 977–1004.

Hawaii Book and Music Festival (HBMF). 2011. *The Value of Hawai'i: Shaping the Future*. Hawaii Council for the Humanities.

Hawai'i Island Journal. 2011. 'Aloun Farms forced labor case goes to trial.' http://www.hawaiiislandjournal.com/2011/07/aloun-farms/ (accessed 15 June 2015).

Hollier, D. 2014. 'Can Hawai'i feed itself.' *Hawai'i Business Magazine*. November 2014. http://www.hawaiibusiness.com/can-Hawai'i-feed-itself/ (accessed 15 June 2015).

Johnson, D., N. Oyama, L. LeMarchand and L. Wilkens. 2004. 'Native Hawaiians mortality, morbidity, and lifestyle: Comparing data from 1982, 1990, and 2000.' *Pacific Health Dialog* 11(2): 120–130.

Kēhaulani, K., T. Lomawaima, F. Mallon, A. Ramos and J. Rappaport. 2008. *Hawaiian Blood: Colonialism and the Politics of Sovereignty and Indigeneity*. Durham, NC: Duke University Press.

Koberstein, P. 2014. 'GMO companies are dousing Hawaiian island with toxic pesticides.' Grist.org. http://grist.org/business-technology/gmo-companies-are-dousing-hawaiian-island-with-toxic-pesticides/ (accessed 15 June 2015).

Krystallis, A. and M. Ness. 2005. 'Consumer preferences for quality foods from a South European perspective: A conjoint analysis implementation on Greek olive oil.' *International Food and Agribusiness Management Review* 8(2): 62–91.

Leong, J., J. Marra, M. Finucane, T. Giambelluca, M. Merrifield, S. Miller, J. Polovina, E. Shea, M. Burkett, J. Campbell, P. Lefale, F. Lipschultz, L. Loope, D. Spooner and B. Wang. 2014. 'Hawai'i and U.S. Affiliated Pacific Islands. Climate Change Impacts in the United States.' In *The Third National Climate Assessment*, edited by J. Melillo, Terese (T.C.) Richmond and G. Yohe. U.S. Global Change Research Program, pp. 537–556.

Liu, D. and C. Alameda. 2011. 'Social determinants of health for Native Hawaiian children and adolescents.' *Hawai'i Medical Journal* 70 (11 Suppl 2): 9–14.

Louis, E. 2012. *The Political Ecology of Food Sovereignty Movements in Neoliberal India*. Doctoral dissertation, University of Hawai'i at Manoa.

McCarthy, J. 2006. 'Rural geography: Alternative rural economies: The search for alterity in forests, fisheries, food, and fair trade.' *Progress in Human Geography* 30(6): 803–811.

McIntyre, L. and K. Rondeau. 2011. 'Individual consumer food localism: A review anchored in Canadian farmwomen's reflections.' *Journal of Rural Studies* 27(2): 116–124.

MacLennan, C. 2014. *Sovereign Sugar: Industry and Environment in Hawai'i*. Honolulu: University of Hawai'i Press.

Mostafanezhad, M., K. Suryanata, S. Azizi and N. Milne. 2015. '"Will weed for food": The political economy of organic farm volunteering in Hawai'i.' *Geoforum* 65: 125–133.

Onozaka, Y. and D. McFadden. 2011. 'Does local labeling complement or compete with other sustainable labels? A conjoint analysis of direct and joint values for fresh produce claim.' *American Journal of Agricultural Economics* 93(3): 693–706.

Parker Ranch 2014. "Paniolo cattle company formed by Parker Ranch and Ulupono Initiative." Press Release, Parker Ranch, Hawai'i.

Roitner-Schobesberger, B., I. Darnhofer, S. Somsook and C. Vogl. 2008. 'Consumer perceptions of organic foods in Bangkok.' *Thailand Food Policy* 33(2): 112–121.

Seyfang, G. 2006. 'Ecological citizenship and sustainable consumption: Examining local organic food networks.' *Journal of Rural Studies* 22(4): 383–395.

Sirieix, L., P. Kledal and T. Sulitang, 2011. 'Organic food consumers' trade-offs between local or imported, conventional or organic products: A qualitative study.' *Shanghai International Journal of Consumer Studies* 35(6): 670–678.

Suryanata, K. 2002. 'Diversified agriculture, land use, and agrofood networks in Hawai'i.' *Economic Geography* 78(1): 71–86.

Thøgersen, J. and Y. Zhou. 2012. 'Chinese consumers' adoption of a "green" innovation: The case of organic food.' *Journal of Marketing Management* 28(3–4): 313–333.

US Census Bureau 2010. *2010 Census Demographic Profile*. www.census.gov/prod/cen2010/doc/dpsf.pdf (accessed 14 June 2015).

Voon, J., K. Ngui and A. Agrawal. 2011. 'Determinants of willingness to purchase organic food: An exploratory study using structural equation modeling.' *International Food and Agribusiness Management Review* 14(2): 103–120.

Wang, Q. and J. Sun. 2003. 'Consumer preference and demand for organic food: Evidence from a Vermont survey.' Conference Proceedings of the American Agricultural Economics Association Annual Meeting, Montreal, Canada, 27–30 July.

6 The food revolution in Melbourne, 1980–2015

Warwick Frost and Jennifer Laing

In 2014, Tourism Australia launched its Restaurant Australia marketing campaign to great fanfare. Journalists, chefs and celebrities from around the world, including the likes of A.A. Gill, Gwyneth Paltrow, Heston Blumenthal and Alice Waters, were invited to various media events that promoted Australia as a top culinary tourist destination. As Tourism Australia noted in their fact sheet: 'Australia's food and wine has become one of its greatest assets with a range and quality of produce that is second to none. Yet, remarkably, the appeal of Australian food and wine remains one of our best-kept secrets' (Tourism Australia 2015). While this might be true of some parts of Australia, it ignores the steady development of Melbourne as a foodie city and destination since the early 1980s.

Melbourne's culinary excellence is acknowledged on the global stage. In 2014, it was named number seventeen of the world's eighteen food cities by Thrillist Media Group. The criteria for selection was 'uniqueness of traditional food/drink style, quality of restaurants/bars, diversity of cuisines, and also, for the most part, a feeling that it is changing and improving' and Melbourne was the only Australian city to make the list (House 2014). In addition, Attica restaurant in suburban Melbourne was recently voted number thirty-two in the world and was the only Australian restaurant to make the list of the World's Top 50 restaurants, 'started in 2002 by British *Restaurant* magazine, [which] is the most respected and anticipated restaurant hit-list of its type internationally. The top 10 particularly becomes an instant dining bucket list for globetrotting food lovers' (Bernoth 2015). While Melbourne doesn't boast any Michelin-starred restaurants, this might be a function of the inherent conservatism of the Michelin rating system, including its bias against national and ethnic cuisines in favour of 'French cultural hegemony' (Lane 2013: 362). Complementing this gastronomic recognition, in 2015 Melbourne was named the World's Most Liveable City by the Economist Intelligence Unit for the fifth consecutive year (Economist Intelligence Unit 2015; Holden and Scerri 2013).

The last thirty-five years or so has seen a dramatic shift in Melbourne's culinary culture. The city now sees itself as a major gastronomic centre with a vibrant restaurant and café culture. An appreciation of food, particularly the latest trends in gastronomy, is seen as contributing to a sophisticated and modern self-identity. While important to residents of the city, this change is also a significant component of a successful tourism strategy for Melbourne.

In this case study, we explore these changes using a variety of historical and contemporary sources, along with qualitative interviews we conducted with city officials. Our aim is to examine some of the explanations of how and why this transformation took place over the last thirty-five years. We begin with a section that explores how Melbourne became such a foodie city, consisting of a historical overview, a discussion of the city's competitive advantage and a summary of the critical policy and regulatory changes. We follow this with an examination of the role of diaspora communities in shaping food consumption practices, trends regarding food consumption at home and the environmental implications of Melbourne's food needs.

Melbourne, the foodie city

Historical overview

Historically, the food economy of Melbourne has been quite different from those of most other cities and regions in the Asia-Pacific area. A European settlement established in 1834 very quickly led to the displacement and near extermination of the Indigenous peoples. Wrongly imagining it as *Terra Nullius* (an empty land), the newcomers imported an entrepreneurial society based on commercial large-scale production focused on European mass markets. Unlike Sydney half a century earlier, there was no potential for starvation, as the new settlement was fully integrated into European trading networks.

The discovery of gold in 1851 stimulated economic development and rapid population growth; 'the new city expanded furiously ... it was an extraordinary boom, yet in tune with the peculiarly urban nature of the British Empire in Australia' (Hunt 2014: 308). By the late nineteenth century, Melbourne was seen as having one of the highest per capita incomes in the world and its affluence was demonstrated by wide streets, gothic public buildings and extensive parks and gardens (Hunt 2014). The expanding agricultural frontiers into dryland wheat farming, irrigated crops and dairying not only easily fed the booming metropolis, but more importantly drove a major food exporting economy – albeit with dramatic environmental implications (Frost 1997, 1998, 2002a, 2004; Tyrrell 1999). Melbourne was a prime example of *staple theory* – once a popular concept, but less used nowadays by economic historians – in which economic development is due to a small number of highly influential export commodities. In Melbourne's case these were gold, wheat, wool and butter. Similar patterns were in evidence in other Pacific settler economies, such as Auckland, Dunedin and San Francisco; but importantly Melbourne's food economy was very different to that of rapidly growing Asian cities like Singapore and Hong Kong.

Melbourne's history, then, has been of a wealthy city with access to a great deal of cheap food. Until recently, this manifested itself in high consumption, particularly of fats and proteins. Gastronomically, until the late twentieth century, Australians ate quantity rather than quality (Symons 2007). As shown in Table 6.1, food consumption in Australia in the second half of the twentieth century was greatly focused on protein. Total consumption of meat has sat at around 100–110 kg per person per year, or just under 300 g per day. While beef and lamb has fallen, this has been matched by increases in pork and poultry. This shift is partly due to dietary concerns and partly due to changes in ethnicity (which is discussed later in this chapter). Dairy products have increased slightly, though egg consumption has fallen dramatically, probably due to publicity regarding cholesterol. The major increase in fruit is primarily due to a major surge in fruit juice consumption and this is paralleled by a similar rise in soft drinks. The growth in vegetable consumption is surprising and seems to go against perceived wisdom about the quality of the Australian diet. The increase is most apparent with potatoes and tomatoes, with a slight decline in leafy and green vegetables.

The Australian Bureau of Statistics (ABS) has not produced these statistics since 1999. Instead they have shifted to a survey system, asking a sample of respondents to recall what they ate in the last twenty-four hours (ABS 2014). However, this data is not comparable with the historical statistics used in Table 6.1. Furthermore, the ABS admits that its new survey has serious limitations in respondents under-reporting what they actually ate (ABS 2014). This was not due to poor memory recall, but rather to respondents providing the answers they thought authorities were seeking. Unfortunately, these issues mean that we know far less about food consumption in Australia than we did a decade ago.

Table 6.1 Apparent per capita food consumption for Australia, year ended 30 June.

Food (measure)	1939	1969	1999
Red meat (kg)	98	79	53
Pork (kg)	9	10	28
Poultry (kg)	n/a	8	31
Total meat (kg)	107	97	112
Dairy (kg)	18	25	23
Fruit (kg)	79	87	135
Vegetables (kg)	n/a	124	162
Eggs (no)	243	222	137
Grain products (kg)	141	146	138
Cane Sugar (kg)	51	52	43
Alcohol (litres)	3.4	7.5	7.5
Soft drink (litres)	n/a	47	113

Source: ABS, 2000. Note this data is only available for Australia and not for Melbourne.
Apparent consumption is calculated by adding production and imports and subtracting exports.

Contemporary Melbourne's competitive advantage

Melbourne's food scene enjoys a reputation for eclecticism and innovation, incorporating different price points and a varied and vibrant range of cuisines. While one can eat a degustation menu at Attica or the Fat Duck for AUD 300–500 per person, there are many cheaper alternatives available. In 2015, we interviewed senior managers in the City of Melbourne and one observed:

> The other thing that Melbourne is famous for is that [many of] its restaurants are mid range. So there are interesting restaurants, but they're mid range restaurants, they're not your Michelin and your five chef's hatted restaurants, but the food is usually very good and the experience is an interesting one (R1).

A recent change for Melbourne has been the move away from fine dining towards a more casual dining experience. The city flaunted a number of 'big occasion' restaurants in the 1980s and 1990s, such as Stephanie's, owned by celebrity chef Stephanie Alexander, Paul Bocuse, Mietta's, Jacques Reymond and Florentino's, but the attrition rate was high. In 2015, only the last two named restaurants still remain open, although Jacques Reymond has now handed over his restaurant to his former sous-chefs, who renamed it Woodland House. Shared plates and shared tables are increasingly offered instead of a carefully and exquisitely plated individual dish for a diner and a separate table for two. Table linen is also less common, while silver service is almost non-existent. What matters is authenticity and a place where the diner feels comfortable.

The growth in ethnic cuisines is a key feature of contemporary Melbourne, the result of diasporas who moved to Melbourne in successive waves, notably from Europe, Vietnam, China and India, as well as more recently Africa and the Middle East. While Lane (2013) notes that the Michelin Guide usually confines restaurants with ethnic cuisines to the one-star rating out of three, Melbourne's *The Age Good Food Guide 2016* has awarded two chef hats (the maximum is three) to the Flower Drum and Spice Temple (Chinese), Grossi Florentino Upstairs and Rosetta (Italian), Minamishima (Japanese) and The Press Club (Greek).

Melbourne's coffee culture is possibly its single most characteristic feature. Comparing Melbourne with Vancouver – they had tied as the World's Most Liveable City – Holden and Scerri describe coffee in Melbourne today as:

> Absolutely everywhere, and almost completely unfranchised; it is bleary-eyed dependency and a core social institution; it is precious luxury and banal work-a-day routine. How they manage to convince all the small businesses to purchase expensive espresso machines, learn to pull excellent shots, every time, all across the city, without specific draconian legislation defies explanation ... Melburnians say simply: businesses would go under if they didn't serve a superb macchiato.
>
> (Holden and Scerri 2013: 447)

The failure of the North American franchises to gain a substantial foothold, notably Starbucks (Frost et al. 2010), is a source of pride for Melburnians, as is the reputation of its coffee overseas. In Vancouver, cafés advertise on billboards that they employ 'Melbourne-trained baristas', while the barista at a groovy Nottingham café in the United Kingdom slipped into the conversation while we ordered our café lattes that she had worked in Melbourne; clearly intended as a signature of cultural capital.

Important elements of Melbourne's foodie reputation are its small bars and laneways. The bars provide intimate and private spaces for eating and drinking – the height of urban cool. They have long opening hours and generally offer a more discerning drinks list and a more stylish ambience than the standard hotel or pub, minus the television screen and poker machines. Melbourne's graffiti-covered laneways are hangouts for hipsters in the evening, and popular settings for photographers and tourist groups during the day. With an eye to visitation, a number have been renamed after pop music legends such as AC/DC Lane (Frost 2008) and Chrissie Amphlett Lane (the lead singer of The Divinyls), both akin to Joey Ramone Place in New York. Hosier Lane, now renowned for its extensive street art, attracts tourists in numbers that have surprised city officials (we are currently working on a project documenting and exploring this radical change).

There is a diversity of eating places in these laneways, catering for all tastes and budgets; everything from a 'hole in the wall' coffee provider to Frank Camorra's MoVida in Hosier Lane, awarded one chef's hat in 2016 by *The Age Good Food Guide* and regional Peruvian fare at Pastuso in AC/DC Lane (Figures 6.1 and 6.2). They also make an atmospheric setting for food-related festivals, including the MARCS (Melbourne Art Restaurant Cocktail Sound) Laneway Festival. A representative of the City of Melbourne we interviewed told us why he felt the laneways were so important to Melbourne:

> Melbourne does *small* beautifully. So in Sydney you've got a wonderful Opera House, wonderful bridge, wonderful harbour and stuff like that. Melbourne has this wonderful fine grain layered nuanced element to it and there's that sense of discovery. Now the laneways are perfect in that regard (R1).

The laneways' grungy vibe has been strategically highlighted by Tourism Victoria and the City of Melbourne in their promotion of the city to visitors, through campaigns such as 'It's Easy to Lose Yourself in Melbourne' (2006–2010), featuring a girl finding her way around the city using a ball of string tied to her bedpost. The message is that Melbourne is a place of hidden treasures, rewarding the adventurous tourist who seeks them out. Other Australian capital cities have subsequently tried to copy this bar and laneway promotional strategy. The Adelaide City Council is now upgrading its laneways and has issued almost fifty new small bar licences in the past year (ABC 2015). A newspaper advertisement in May 2015 for the campaign titled 'Adelaide.

Breathe', includes the exhortation, 'Take your time to meander through laneways and alleyways and discover the energy and intimacy of our city's small bar culture', in a clear nod to Melbourne's success. Sydney began its laneway revitalisation program in 2008, involving new paving, lighting and public art installations, and road closures (City of Sydney 2015), and made their small bar scene more viable when they changed their liquor licensing laws the same year to mirror Melbourne's. Despite this, neither city has rivalled Melbourne as a foodie city, perhaps due to other elements discussed in this chapter, which collectively form a distinctive package which is not easily imitated by other destinations.

Melbourne's final point of competitive advantage is that it combines food with other complementary lifestyle activities and experiences, notably retail, sport and the arts. This is reflected in its destination marketing, which is 'increasingly focused on culture, cuisine, popular culture, fashion and lifestyle – a sophisticated modern city rather than one with one or two tangible icons' (Frost et al. 2010: 99). This offering is particularly attractive to the creative class identified by Florida (2002, 2005), who enjoy lives 'packed full of intense, high quality, multidimensional experiences' (2002: 166) and seek authenticity in the places where they eat and in the food they consume.

Figure 6.1 A hole in the wall café with grungy/hipster vibe in Hosier Lane.
Photo by J. Laing.

Figure 6.2 Hosier Lane combines street art, restaurants and cafés and has become an iconic tourist attraction.
Photo by J. Laing.

The policy and regulatory backdrop to Melbourne's foodie renaissance

The catalyst for Melbourne's evolution as a foodie destination is the 1988 reforms to liquor licensing, stemming from a government enquiry headed by economist John Nieuwenhuysen, which allowed its cafés to serve wine and other alcoholic beverages. It was essentially a vision of a European café lifestyle (Frost et al. 2010), and livened up Melbourne's city centre after dark, as well as during the day. In our interview, R1 explained the significance of these reforms to Melbourne's food scene: '[Nieuwenhuysen] brought in the regulations that allowed small businesses and small bars, not big beer bars, but again it was that notion of *small*. And rents were low and this meant that businesses could profit or prosper.'

Two other important individuals behind Melbourne's food renaissance in the mid-1980s were Rob Adams, an architect and planner and now Director of City Design at the City of Melbourne and Professor Jan Gehl, a Danish architect. Adams recognised the urgent need for revitalisation of the urban centre and engaged Gehl to conduct a survey of Public Spaces and Public Life in Melbourne. The findings were set out in a report titled *Places for People: Melbourne City 1994*, and there have been two further versions of this survey leading to reports in 2004 and 2014. This research provided the support for improvements to Melbourne's laneways, the development of outdoor dining areas such as Federation Square, increases in the number and seating capacity of outdoor cafés and improvements to the quality of their furniture; encouraging the trend towards al fresco dining that has become the city's trademark.

The role of diaspora communities

In recent years, researchers have given greater focus to the role of diaspora communities in shaping cuisines and foodways (Frost and Laing 2015; Gabaccia 1998; Laing and Frost 2013; Timothy and Ron 2013). As this research has progressed, it has shaped a fundamental shift in how we see these diaspora communities. In the past, there was a tendency to see them as fossilised relics, where cuisine was preserved as a marker of a lost heritage. Nowadays, interpretations tend to take a more dynamic focus, recognising that cuisines and cultures continue to change and that there is a greater trend towards hybridisation as diaspora and host communities interact.

Melbourne is an important and instructive example of this interaction. Indeed, it is popularly held that Melbourne's culinary transformation is largely dependent on the contribution of waves of migration. Originally, Melbourne's population was primarily British. While the Gold Rushes attracted people from around the world – particularly from China – a greater distance from the Atlantic economies meant that society was less multicultural than the other gold economy of California (Frost 2002a, 2002b). In the last half of the nineteenth century and first part of the twentieth century, migration schemes and institutions were focused on British migration. The introduction of the White Australia Policy at Federation capped a trend towards excluding Asian migrants. Following World War II, there was however a major shift in migration policy. Under the slogan of 'populate or perish', there were great waves of assisted European migration between 1945 and 1970. Then, throughout the late twentieth century, a greater focus on economic and political ties with Asia led to the opening up of migration from that region.

As a prosperous city with the main manufacturing economy in Australia, Melbourne attracted a higher proportion of these migrants. The results can be seen in the population census data (Table 6.2). This shows the percentage of the city's population in the main ethnic groups. This data is compiled by ancestry, a census question developed as a more accurate measure of ethnicity than birthplace. However, some care must be taken with its interpretation. Like all census questions, the answers are provided by the respondent and there is no process of verification. Furthermore, multiple answers are allowed, so that a person could, for example, record that they have Irish, Chinese and Swiss ancestry (Laing and Frost 2013).

The results from the most recent census confirm the perception of Melbourne as a multicultural city. While the predominant ethnicity is British, there are quite sizeable minority groups. The Italians, Greeks, German, Dutch and Maltese are products of the post World War II assisted migration schemes in which the Australian government provided fares and support with employment.

Table 6.2 Population of Greater Melbourne by ancestry

Ancestry	%
Australia	28.3
British (English, Scottish and Welsh)	37
Irish	9.4
Italian	7.5
Chinese	6.5
Greek	4.1
Indian	3.4
German	3.3
Vietnamese	2.1
Dutch	1.7
Maltese	1.7
Polish	1.2
Filipino	1.2
Macedonian	1.1
Lebanese	1

Source: Calculated from ABS, 2015. Multiple responses.

In contrast, the Chinese, Indian, Vietnamese and Filipinos are more recent migrant waves (there is no way to distinguish the Chinese as to whether they came in the nineteenth century or twenty-first century, though the great majority are of recent origin). These statistics are for the whole city, though percentages of ethnic groups are much higher in certain neighbourhoods historically connected with new migrants.

In popular perception, these diaspora groupings are viewed as greatly influencing Melbourne's food culture. Modern evidence for this is seen in the wide array and large numbers of ethnic restaurants and speciality food stores. Many of these are clustered in certain neighbourhoods. For example, the shopping strip of Victoria Street Richmond is dominated by Vietnamese businesses. However, the patterns and influence of these diaspora communities is more complex than in the popular narratives. To explore these more fully, we consider case studies of the Chinese and Italian communities.

The Chinese diaspora

The Chinese came for gold in the 1850s, becoming the largest non-English speaking diaspora of the nineteenth century. Initially, their focus was on gold, but declining yields by the 1860s forced those who remained to engage in other activities. Some began to grow vegetables commercially, working in small teams. Here they were responding to a gap in the market, for European farming tended towards wheat and livestock farmed on a large scale. Labour was short and wages high, so there was an advantage for a migrant group willing to work cooperatively. Around Melbourne, the Chinese dominated vegetable growing in the sandbelt to the south of the city and the alluvial creek valleys to the north. Contrary to modern popular opinion, they did not grow Asiatic crops. Entrepreneurial in their focus, they specialised in the well-known European vegetables that the market demanded (Frost 2002b).

In the city centre, a large Chinatown developed along three blocks of Little Bourke Street. Here were merchants and small restaurateurs. Renowned as a red-light district, Little Bourke Street attracted curious Europeans intent on an edgy night out. Cheap meals added to the experience. As with other cities around the world, such ethnic ghettoes combined the twin appeals of exotic difference and a little bit of danger in stepping outside the everyday (Hall and Rath 2007).

By the twentieth century, Chinese restaurants were found across Melbourne and in surrounding country towns. Advertising Australian and Chinese meals, they sold staples like steak and eggs, but the adventurous would also try fried rice and noodle dishes. Chop suey, invented in San Francisco's Chinatown, made the journey across the Pacific (Pilcher 2006). Some Chinese entrepreneurs went further in introducing Chinese-style manufactured foods. After World War II, dim sims became popular as a snack sold by corner stores and takeaways. Based on steamed dumplings, they were larger, had thicker pastry and were deep-fried. Spring rolls also got a makeover, repackaged for Australian tastes as Chiko rolls. Like dim sims, these had thick pastry and were deep fried.

As restaurant culture boomed in the 1980s, Chinese restaurateurs adapted and thrived. BYO (Bring Your Own) alcohol licensing was perfect for these family operations. Patrons could bring their own drinks and the operators had no need to keep expensive stocks of unfamiliar wine. When Vietnamese food became highly popular, many of the chefs and restaurateurs were Chinese. Indeed, even today, most Vietnamese restaurants have both Vietnamese and Chinese menus.

The Italian diaspora

Directly after World War II, large waves of Italian assisted migrants came to Melbourne. They quickly concentrated in Lygon Street Carlton. Before the War, this area had been largely Jewish and it had a reputation for being tolerant, arty and bohemian. The Italian newcomers complemented this image. Especially at night and the weekends, they clustered around cafés organised along regional lines. Here, the new migrants could hear news from home and maintain community networks. Before long, they were attracting non-migrants, drawn by a vibrant culture and the lure of something different (Frost and Laing 2015; Hall and Rath 2007).

Espresso coffee was at the heart of this experience. The popular view – reinforced by modern-day media, museum displays and interpretation – is that espresso machines were introduced by the Italians. An alternative version is that it was Peter Bancroft, a young Melbourne man, who had witnessed the popularity of late night cafés while living in London in the 1950s. With his father, he purchased the rights to import Gaggia coffee machines from Italy. They opened a café called Il Cappuccino for demonstration purposes. By 1957, they had sold 400 Gaggia machines. Most were purchased by new migrants. They had received assisted passage to come to Melbourne to work in factories and construction, but many were from a small business background and their dream was to work for themselves (Brown-May 2001). Accordingly, Italian cafés began to spring up around the city and inner suburbs.

Italian cafés and restaurants sold more than pasta and pizza. What also attracted non-Italians was a sense of *festa* and family. This was a different way of eating and enjoying oneself (Frost and Laing 2015). Intriguingly, what was served was quite different from what people ate in Italy. This was a hybridised cuisine, richer and with more protein. As in the USA, Italian migrants found that their incomes had risen and they could afford to routinely indulge in dishes that were usually reserved only for holidays (Gabaccia 1998; Pilcher 2006).

The Anglo-Saxon majority enthusiastically embraced this Italian experience. Three main patterns are particularly apparent. The first is that good-quality meals could be had cheaply in convivial surroundings. The second is that there was a culture of small family-operated restaurants and cafés, often with an owner-chef and their family serving. In contrast to North America, chain operations did not become the norm (Holden and Scerri 2013). Third, a strong coffee culture took hold. Based on independent owner-operators, Melbourne's coffee scene has become a major part of its tourism imagery and US chain stores such as Starbucks failed in the 2000s to have much impression on the market (Frost et al. 2010).

Eating at home

So far, much of our discussion has concerned eating in commercial restaurants and cafés. What about food consumption patterns at home? Three academic studies undertaken in the 2000s have provided an insight into the domestic food habits of people living in Melbourne and linked the findings to the socio-economic status (SES) and the socio-economic position (SEP) of households. Dixon and Banwell (2004) used focus group data collected in Melbourne in 1996 and 1997 to explore the influence of children on food-related decisions. This data is therefore the least recent of the three studies, but may still aid our understanding of modern family food consumption habits in Melbourne. Thirty-three households in total took part in five focus groups and the SES of each household was noted, to see if this affected domestic food consumption. In most cases (twenty-seven families), the women played a major part in shopping for and preparing meals for their families and saw this as 'an accepted and important part of their lives' (Dixon and Banwell 2004: 189), although not necessarily pleasurable. Eating at restaurants and buying take-away food didn't occur on a regular basis for these households, but rather as occasional treats. Mostly this was due to the cost of eating out. However 'when fast food does replace home cooking for lower SES groups the food is consumed in a fast food restaurant' (Dixon and Banwell 2004,:187). Children were increasingly involved in decisions about the food to be consumed, partly as a means of avoiding conflict. Concerns were raised in this study that this might lead to a long-term increase in obesity levels, with children more vulnerable to advertising that encourages bad eating habits.

Campbell et al. (2002) examined the family environments of five- to six-year-old children living in Melbourne and Geelong (the second biggest city in Victoria, after Melbourne), to see how this affects food intake and whether this differs according to SES. Questionnaires were completed and returned by 560 families. The findings show that almost half of the families found it difficult to eat breakfast together, while one-third of families reported difficulties in having dinner together. In both cases, work commitments were cited as the reason for this state of affairs. The education of the mother was a factor in some of the food environments. Families with mothers with low levels of education were more likely to eat together, but also to watch television while they ate. These families also ate more fast food/takeaway food and less fresh fruits and vegetables. Maternal education was not however a factor when considering involvement of children in food preparation. It was found that 45 per cent of the children studied either never took part in preparing food for the evening meal, or only did so less than once a month. The study paints a bleak picture of the environment within which many children in Melbourne experience food and learn to make various food choices. It was noted that these findings 'are consistent with international research, however, to our knowledge, this is the first Australian study to highlight disparity of food environments in non-rural settings' (Campbell et al. 2002: S560).

In the third study by MacFarlane et al. (2007), there was further examination of the potential relationship between the domestic food environments of adolescents in Melbourne and its environs and the SEP of their families. Survey data were collected from 3264 adolescent students and the parent who mainly handled meal planning and preparation in the household. Some findings differed along SEP lines. Thus:

> The home meal environments of high SEP adolescents were more conducive to healthy eating, with adolescents of more highly educated mothers more likely to report that vegetables were always served at dinner, that the evening meal was never an unpleasant time for their family, and that the evening meal was always or usually a time when their family really talked and caught up with each other.
>
> (MacFarlane et al. 2007: 752)

Care should be taken with such results. As noted with the recent ABS Health Survey (ABS 2014), there may be a tendency for respondents to report food consumption patterns that make them look good, emphasising a healthy diet that is not really occurring. Further research would be needed to examine this phenomenon in more depth. This study also highlighted the importance of interventions targeting Australian homes, rather than just schools, to improve 'the home availability of healthy food ... and encouraging the home meal environment to be supportive of healthy eating' (MacFarlane et al. 2007: 754).

Environmental aspects of food consumption

Melbourne draws its food supply from two main areas. The first is distant production regions. Major food products transported large distances include wheat, rice, sugar and tropical products such as fruit, coffee and cocoa. Such disconnection between production and consumption creates a *ghost acreage* of food supply, a term used in cases where affluent cities draw their food from colonial or less developed regions. This does mean that political and environmental conditions elsewhere may have an impact on what the people of Melbourne consume. For example, twice in the last ten years cyclones have destroyed the banana crops in northern Australia. With high tariffs on imported bananas, this has led to very high prices lasting for a year. Having production occurring elsewhere transfers most environmental issues to distant places and it may be that Melbourne consumers are unaware of the impacts they are having. The exception is that there are clear environmental costs in transporting food into Melbourne and these are visible in traffic congestion and pollution.

The second production area is the productive peri-urban fringe. Preliminary results of a recent study argue that the area within a 100 km radius of the city centre produces about 40 per cent of the food needed to supply Melbourne. This includes about 80 per cent of fresh vegetables, plus poultry, red meat, eggs and fruit (Carey et al. 2015). Such agricultural production is only viable on the city fringe as it meets the demand for fresh, high-value produce. It is, for example, perfectly possible to grow wheat just outside Melbourne. However, no farmer does this as the price of the land does not justify producing a bulky low-value grain that is easily transported without any ill-effects on quality. Instead it makes more economic sense to grow high-value perishable produce close to the market.

The key environmental problem in this peri-urban zone is the competition from housing. In 2015, Melbourne's population is 4.4 million and it is estimated to reach 7 million by 2050. As the city grows, it will not only demand more food, but consume the land that is currently producing food. Accordingly, the percentage of food supply produced within 100 km of Melbourne is estimated to fall from 40 per cent in 2015 to 18 per cent in 2050 (Carey et al. 2015). As the population grows and agricultural activities are pushed further outwards, this is likely to increase the carbon footprint of transporting food into Melbourne.

Issues of conflict between farming and suburban growth are already apparent. In a number of cases, municipal councils have been pressured into introducing regulations that restrict commercial farming on the basis of noise, smell, visual amenity and truck access. A further concern has been the largely unplanned spread of housing and hobby farms into the hill country to the north and east of the city. In an environment dominated by dense eucalyptus forests, this has greatly increased the risk and impact of bushfires. In the 2009 Black Saturday Bushfire, 173 people died and over 2000 houses were destroyed.

A third area is worthy of consideration, though it has not been the subject of any research studies and no empirical data is available to gauge its size or impact. Residents of Melbourne have traditionally produced vegetables and fruit from home gardens. In recent decades, this production has seemingly increased, though no data is collected for non-commercial production. Anecdotal evidence indicates that it is possible to be almost self-sufficient in vegetables and fruit from a home garden, though it is likely that most people are not working at that level of intensity.

Interest in home production takes both mainstream and alternative forms. A range of lifestyle programmes on television regularly provide advice on this topic — and these are noticeably sponsored by large hardware and gardening supplies stores. The shift to home production is probably due to two reasons. The first is a desire for better quality organic food (Frost et al. 2016). The second is that it helps people acquire a sort of cultural capital. In the past, gaining cultural capital from food has been linked to travel and consumption (Richards 2002). This new iteration, however, sees people as gaining status and satisfaction through production and the conspicuous display of that production. As well as growing food plants, this trend is also manifested through home artisanal production, with examples including baking one's own bread or making yoghurt. While difficult to quantify, it is likely that home production will continue to grow in the future.

Conclusion

A product of nineteenth-century European settlement of Australia, from its beginning Melbourne was fully integrated into global agricultural and food economies. Accordingly, there was no traditional food phase and very little was adapted from Aboriginal culture. Instead, a successful export economy provided the wealth to stimulate local farming and food imports. With high relative incomes, the residents of the city maintained a diet that was high in calories, albeit one based on the quantity of consumption rather than the quality of cuisine.

The last thirty years has seen a major shift in food consumption patterns in Melbourne. Partly due to changes in government regulation and partly due to the legacy of waves of migration, Melbourne has been transformed into a leading culinary city. In combining fine dining restaurants with affordable cafés, Melbourne's image has become one of a sophisticated foodie metropolis. This has not only contributed to attracting greater flows of tourism, but has led to a strong and positive self-identity for its residents. Success in being named the World's Most Liveable City has validated these perspectives.

Nonetheless, despite these changes, two qualifications need to be acknowledged. The first is that rapidly increasing population growth and urban sprawl are reducing food production on the city's peri-urban fringe. In the long run this will increase environmental externalities (especially through greater transport) and lead to rising food prices. Second, recent research into home food consumption paints a more negative picture, particularly concerning fast food and a lack of fruit and vegetables in everyday diets. This raises a potential paradox regarding a sophisticated and varied restaurant culture against poor-quality/high-calorie home consumption.

References

ABC. 2015. 'Adelaide City laneways revival pushes ahead with rundle mall upgrade all but finished.' *ABC Adelaide*. http://www.abc.net.au/news/2015-07-16/adelaide-laneways-council-twin-street-gawler-place/6624264 (accessed 18 August 2015).

ABS. 2000. *Apparent Consumption of Foodstuffs, 4306.0, 1997–98 and 1998–99*. Canberra: Australian Bureau of Statistics.

ABS. 2014. *Australian Health Survey: Nutrition First Results – Food and Nutrients, 4364.0, 2011–12*. Canberra: Australian Bureau of Statistics.

ABS. 2015. *2011 Census of Population and Housing, Greater Melbourne: Table X06 Ancestry*. Canberra: Australian Bureau of Statistics.

Bernoth, A. 2015. 'Attica voted Australia's top restaurant at World's 50 Best Awards 2015.' *Good Food*, 1 June 2015. www.goodfood.com.au/good-food/food-news/attica-voted-australias-top-restaurant-at-worlds-50-best-awards-2015-20150602-ghdpa4.html (accessed 18 August 2015).

Brown-May, A. 2001. *Espresso! Melbourne Coffee Stories*. Melbourne: Arcadia.

Campbell, K., D. Crawford, M. Jackson, K. Cashel, A. Worsley, K. Gibbons and L. Birch. 2002. 'Family food environments of 5-6 year-old-children: Does socioeconomic status make a difference?" *Asia Pacific Journal of Clinical Nutrition* 11(Suppl. 3): S553–S561.

Carey, R., J. Sheridan and K. Larsen. 2015. 'To feed growing cities we need to stop urban sprawl eating up our food supply.' *The Conversation*, 26 October 2015. http://theconversation.com/to-feed-growing-cities-we-need-to-stop-urban-sprawl-eating-up-our-food-supply-49651 (accessed 28 October 2015).

City of Sydney. 2015. 'City of Sydney', *Laneway Revitalisation*. www.cityofsydney.nsw.gov.au/vision/better-infrastructure/streets-and-public-places/laneway-revitalisation (accessed 18 August 2015).

Dixon, J. and C. Banwell. 2004. 'Heading the table: Parenting and the junior consumer.' *British Food Journal* 106(3): 182–193.

Economist Intelligence Unit. 2015. 'The World's Most Liveable Cities.' www.economist.com/blogs/graphicdetail/2015/08/daily-chart-5 (accessed 20 August 2015).

Florida, R. 2002. *The Rise of the Creative Class: And How It's Transforming Work, Leisure, Community and Everyday Life*. Cambridge, MA: Basic Books.

Florida, R. 2005. *Cities and the Creative Classes*. London and New York: Routledge.

Frost, W. 1997. "Farmers, government and the environment: The settlement of Australia's 'Wet Frontier', 1870–1920." *Australian Economic History Review* 37(1): 19–38.

Frost, W. 1998. "European farming, Australian pests: Agricultural settlement and environmental disruption in Australia, 1800–1920." *Environment and History* 4(2): 129–143.

Frost, W. 2002a. 'Powerhouse economies of the Pacific: A comparative study of gold and wheat in nineteenth century Victoria and California.' In *Studies in Pacific History: Economics, Politics, and Migration*, edited by D. Flynn, A. Giráldez and J. Sobredo, pp. 61–74. Aldershot and Burlington: Ashgate.

Frost, W. 2002b. 'Migrants and technology transfer: Chinese farming in Australia, 1850–1920.' *Australian Economic History Review* 42(2): 113–131.

Frost, W. 2004. '*Australia Unlimited*? Environmental debate in the age of catastrophe, 1910–1939.' *Environment and History* 10(3): 285–303.

Frost, W. 2008. "Popular culture as a different type of heritage: The making of AC/DC Lane." *Journal of Heritage Tourism* 3(3): 176–184.

Frost, W. and J. Laing. 2015. 'Cuisine, migration, colonialism and diasporic identity." In *Heritage Cuisines*, edited by D. Timothy, pp. 37–52. London and New York: Routledge.

Frost, W., J. Laing, F. Wheeler and K. Reeves. 2010. 'Coffee culture, heritage and destination image: Melbourne and the Italian model.' In *Coffee Culture, Destinations and Tourism*, edited by L. Jolliffe, pp. 99–110. Clevedon: Channel View.

Frost, W., J. Laing, G. Best, K. Williams, P. Strickland and C. Lade. 2016. *Gastronomy, Tourism and the Media*. Bristol: Channel View.

Gabaccia, D. 1998. *We Are What We Eat: Ethnic Food and the Making of Americans*. Cambridge, MA: Harvard University Press.

Hall, C. and J. Rath. 2007. 'Tourism, migration and place advantage in the global cultural economy.' In *Tourism, Ethnic Diversity and the City*, edited by J. Rath, pp. 1–24. London and New York: Routledge.

Holden, M. and A. Scerri. 2013. 'More than this: Liveable Melbourne meets liveable Vancouver.' *Cities* 31: 444–453.

House, A. 2014. 'Melbourne named among world's top 18 foodie cities.' *Herald Sun*, Travel. http://www.heraldsun.com.au/travel/australia/melbourne-named-among-worlds-top-18-foodie-cities/story-fnjjv1f4-1227085987184 (accessed 18 August 2015).

Hunt, T. 2014. *Ten Cities that Made the Empire*. London: Allen Lane.

Laing, J. and W. Frost. 2013. 'Food, wine … heritage, identity? Two case studies of Italian diaspora festivals in regional Victoria.' *Tourism Analysis* 18(2): 323–334.

Lane, C. 2013. 'Taste makers in the 'fine dining' restaurant industry: The attribution of aesthetic and economic value by gastronomic guides.' *Poetics* 41: 342–365.

MacFarlane, A., D. Crawford, K, Ball, G. Savige and A. Worsley. 2007. 'Adolescent home food environments and socioeconomic position.' *Asia Pacific Journal of Clinical Nutrition* 16(4): 748–755.

Pilcher, J. 2006. *Food in World History*. New York and Oxford: Routledge.

Richards, G. 2002. 'Gastronomy: An essential ingredient in tourism production and consumption?' In *Tourism and Gastronomy*, edited by A.-M. Hjalager and G. Richards, pp. 3–20. London and New York: Routledge.

Symons. M. 2007. *One Continuous Picnic: A Gastronomic History of Australia*. Melbourne: Melbourne University Press.

Timothy, D. and A. Ron. 2013. 'Understanding heritage cuisines and tourism: Identity, image, authenticity and change.' *Journal of Heritage Tourism* 8(2–3): 99–104.

Tourism Australia. 2015. *Restaurant Australia: Key Messages.* www.tourism.australia.com/documents/Campaigns/Restaurant_Australia_key_messages.pdf (accessed 18 August 2015).

Tyrrell, I. 1999. *True Gardens of the Gods: Californian-Australian Environmental Reform, 1860–1930.* Berkeley: University of California Press.

7 The practices of Indian vegetarianism in a world of limited resources

The case of Bengaluru

Rune-Christoffer Dragsdahl

One of the most unknown factors in future scenarios on how to feed the world is the food practices of youth in India. As the country is experiencing a 'Green Revolution fatigue' with stagnant yields and falling water tables (Rowden 2011: 11), the population is expected to rise from 1.3 billion to 1.7 billion by 2065 (United Nations Population Division 2012). These trends are accompanied by a crucial question: what will be the future level of meat consumption in this country?

It is a common perception that anywhere in the world economic growth leads to increased levels of meat consumption, a question debated in Chapter 1 by Hellmuth Lange, in relation to China and India specifically. The basis for this is usually a chart from the Food and Agriculture Organization (FAO), where countries are distributed along a curve showing a positive correlation between increased incomes and increased consumption of meat per capita, although the increase does level off among the most wealthy countries (FAO 2009: 12). However, there are quite a few outliers on the chart, which do not support the assumption of a mechanical correlation. Indeed, India is one such exception from the global pattern, with per capita meat consumption at 5 kg per year (FAO 2009: 138). Without comparison, India therefore also represents the country in the world with the largest potential future increase in meat consumption, which would have significant impacts on food resources and environmental sustainability.

Based on ethnographic material on the food consumption practices of youth in Bengaluru collected in 2009 and 2010,[1] this chapter explores the practices of vegetarianism and non-vegetarianism as negotiated everyday by youth in contemporary Bengaluru. College youth were chosen as the subjects of the research because they are the consumers of food of both today and tomorrow. Furthermore, as my research shows, they are positioned between family-based caste society and an egalitarian society based more on friendship facilitated by life on campus. Bengaluru was chosen as the site of research as changes in food patterns normally happen faster in cosmopolitan settings. If a change towards more meat is underway, this should be more visible in a place like Bengaluru.

My key informants were all Hindus, either current or past college students between nineteen and twenty-three years old. Their names have all been changed to keep their real identities anonymous. They were chosen from a preliminary survey based upon families with at least one family member's food practice deviating from the rest of the family. I also conducted a final quantitative survey, which confirmed the relevance of the key qualitative findings. The survey with 1000 respondents was carried out at ten different college campuses. The respondents were all current college students, representing all religious, caste and social backgrounds, as India has a caste-based quota system for accessing higher education, which creates a very heterogeneous student community.

In my analysis, I particularly emphasize the importance of egalitarian friendships and peer pressure in making young people from vegetarian families start eating meat, the ambiguous practices surrounding the consumption of eggs, and the complex relationship between caste and vegetarianism. I argue that youth in Bengaluru practise an ambiguous balancing of, on the one hand, influences from their family and caste, and on the other hand, influences from their friends and romantic relationships, which result in contextual meat consumption (Caplan 2008), 'selective vegetarianism' (Donner 2008) and ultimately doubting the clear division between vegetarian and non-vegetarian food items. These practices and doubts can all be seen as ways of mediating contrasts between vegetarian and non-vegetarian food, and contrasts between egalitarian relationships and caste society.

Brief review of the literature on veg–non-veg in India

In the following, I will briefly situate my argument in the previous academic literature on vegetarianism and non-vegetarianism in India.

Annual meat consumption levels for most countries in the Global North hover between 80 and 120 kg per capita (FAO 2009). In Euro-American thought, meat consumption is often associated with development (Rifkin 1992), evolution (Peace 2008), status (Twigg 1984) and human power over nature (Fiddes 1991). Back in the nineteenth century, European intellectuals argued that the food hierarchy from plants to red meat corresponded to an evolutionary hierarchy from the wild and coloured races to the white and civilized races (Rifkin 1992: 247). The sociologist Julia Twigg has suggested a similar hierarchy for the status of food categories in Western society, where read meat is on top, followed by fowl and fish, then eggs and dairy products, and at the bottom the low-status foods: fruits, vegetables and grains (1984: 21).

That said, since colonial times India has been allowed a more positive place in Euro-American perceptions of vegetarianism, associated with ayurvedic diets, the philosophy of Gandhi and popular images of 'the sacred cow'. These ideas are not unfounded: the distinction between vegetarian food (*veg*) and meat (*non-veg*) is omnipresent in India, and the negative definition of meat is inherent in the concept *non*-veg. Forty per cent of the population of India are vegetarians (Yadav and Kumar 2006), most of the remaining population only eat meat occasionally, and from 1995 to 2005 annual meat consumption grew very little – from 4.7 kg to 5.1 kg per capita (FAO 2009: 138). More recent data indicate that beef consumption is falling, while chicken consumption is increasing (Raghavan 2015). Overall, meat consumption per capita is still among the lowest in the world.

In *The Eternal Food,* R.S. Khare (1992), one of the leading scholars on the cosmological meanings of food in India, writes:

> Foods, firmly grounded in moral ideals and practice, represent a cosmic, divine principle at one level and a most immediate and intimate semiotic reality at another ... Food in India is never merely a material substance of ingestion, nor only a transactional commodity. It is synonymous with life and all its goals, including the subtlest and the highest.
>
> (Khare 1992: 1–2)

Indeed, in his classic work *Homo Hierarchicus,* Louis Dumont briefly touches upon the subject of vegetarianism as part of his famous argument that caste is 'a state of mind' (1966: 34), which can be reduced to one principle: that pure and impure are opposites that must be kept apart (1966: 43). According to Dumont, vegetarianism is acknowledged by the entire Hindu population as 'the superior food practice' (1966: 146), and he explicitly connects vegetarianism with purity: 'No doubt it must be stated that vegetarianism was by its nature fit to be easily integrated into the ideas about pure and impure' (Dumont 1966: 150). He also notes how the cow divides the upper castes from the lower castes, in the sense that Dalits worked with dead cows, for example through making their skin into leather, while Brahmins are associated with cows (1966: 54).

Since then, Dumont has been heavily criticized for his analysis favouring and promoting what is an ideology of some, but certainly not all, Indians (see for example Das 2006; Appadurai 2006). The same criticism applies to the analysis of the meanings of vegetarianism in India. Much of the emphasis on the positive aspects of vegetarianism probably derives from the dominance of largely vegetarian castes such as well-educated Brahmins in representing India abroad. In recent decades, the idea of India as a vegetarian society has therefore increasingly been contested by scholars such as Chigateri (2008), Dolphijn (2006) and Osella (2008).

In 2008, a special issue of the journal *South Asia: Journal of South Asian Studies* was published. The theme was *Food, Memory, Pleasure and Politics,* and the articles were based on several recent ethnographic studies of food in India. The authors all take their point of departure in the veg–non-veg dualism, but in two rather different ways. Some of the authors (for example, Chigateri 2008; Osella 2008) analyze veg–non-veg as a fundamentally suppressing core of the caste system, while other authors (for example, Caplan 2008; Staples 2008; Donner 2008; Desai 2008; Michelutti 2008; Klein 2008) focus on how veg–non-veg when practised is characterized by complexity, contextuality and paradoxes.

Shraddha Chigateri conducted fieldwork among Dalits in Bengaluru, and she notes how the champion of the Dalits, Ambedkar, argued that beef consumption was historically the root cause for the phenomenon of untouchability in India (Chigateri 2008: 21). Ambedkar distinguished between vegetarian (*shakahari*) and meat-eating (*mansahari*) groups, with a further subdivision of the latter category between those eating beef and those not (Chigateri 2008: 20). Chigateri therefore argues that among Hindus in India, there is a hierarchy from vegetarianism over meat consumption to beef consumption (Chigateri 2008: 11). This hierarchy is not only implicitly violent, as in the case of untouchability; it is also explicitly violent when buildings are incinerated and slaughters harassed or threatened (Chigateri 2008: 15–16). Some scholars, and many Dalits, therefore see the Brahmin food hierarchy as a sort of 'spiritual fascism' (Ilaiah in Chigateri 2008: 21). This may also explain why meat consumption today for some lower castes, according to Rick Dolphijn (2006), has become a radical manifestation of political liberation – a counter-reaction against the symbolic subordination of meat consumers by powerful upper castes.

Chigateri does acknowledge that the food hierarchy in practice is more 'messy' with regional differences, variations between groups and a difference between ideology and practice (2008: 20). However, that complexity is mostly set aside by her emphasis on beef and the veg–non-veg dualism. Caroline Osella even argues that the veg–non-veg dualism can never be separated from caste society and its purity-based hierarchies (Osella 2008: 5) because in India there is 'an always-implicit understanding that vegetarianism articulates here with ideas about Hindu ritual purity, caste pollution and social status' (Osella 2008: 7).

Other authors, however, make it their key point that the veg–non-veg ideology is practised, in reality, in diverse and complex ways. Pat Caplan (2008), who studied the middle class in Chennai, criticizes the kind of essentialism where vegetarianism is associated with Brahmins, and meat consumption with non-Brahmins: many meat eaters view vegetarian food in a positive light, and male Brahmins often eat vegetarian at home and meat outside, or share vegetarian food with meat eaters. Caplan thereby shows that there is a complexity and context-dependent flexibility in the social relations of vegetarians and meat eaters, and this results in an ambivalent perspective on meat, which is sometimes seen as polluting, however not always, while the meat eaters do not mind eating vegetarian *most of the time*, as long as this is not *always*. A similar complexity is seen in Henrike Donner's (2008) study of 'selective vegetarianism' among the middle class in Kolkata.

James Staples even argues that any analysis of food in India based on binary oppositions is problematic (2008: 54). I do not wish to take the argument that far because according to my ethnographic material from Bengaluru, the food practices of youth are permeated by binary oppositions. I find it difficult to deny that given the omnipresence of the green 'veg' sign on everything from fast food restaurants to labels on processed foods. The point I want to emphasize is rather that the binary opposition between veg and non-veg does exist as an idea that people refer to, but that it is constantly negotiated in the everyday lives of people, and changes meaning from context to context. In the following, I will show how contextual meat consumption (see for example Caplan 2008) and 'selective vegetarianism' (see for example Donner 2008) are very fitting terms for the food practices of youth in Bengaluru.

Between friends and family

Among the 1000 young Indians who responded to my questionnaire, 36 per cent were vegetarians and 64 per cent were non-vegetarians. There was a strong correlation between caste and food practice: 86 per cent of the Dalits eat non-veg, 85 per cent of Vokkaligas eat non-veg. Among Lingayaths 84 per cent eat only vegetarian, while among Brahmins this rate is at 88 per cent.

What these numbers also show, however, is that some people do deviate from the traditional food practices of their caste. Across the span of three generations (grandparents, parents and siblings of the person who filled in the questionnaire), at least one family member deviates from the food practice of the rest of the family in 36 per cent of all families. Conversely, this means that 64 per cent of all these families, even across three generations, have no family members deviating from the family's preference for either eating only veg or also eating non-veg. In short, the veg–non-veg divide has a pattern which *is* strongly caste-based, and yet such caste-based food patterns are commonly challenged when practised.

One of the primary challenges towards traditional, caste-based patterns of meat consumption comes from friendships across caste barriers. Nicholas Nisbett describes how a group of young men from the middle class in Bengaluru negotiate conventional hierarchies by creating an egalitarian youth culture across caste barriers. The community of friendship is their alternative to a caste-based hierarchy (Nisbett 2007: 948). In a similar way, my informant Hemant, twenty-three years old, explained how he sees friendships as the opposite of casteism:

> I don't allow the casteism. I have all kinds of friends. I have Muslim friends, I have Brahmin friends also, I have Kshatriya friends also, I have Lingayath friends also, I have Scheduled Caste friends also. That's why I don't allow that one. I hate that one.

This criticism towards casteism, however, does not imply that youth prioritize friendships higher than their own family and thereby their caste. According to my questionnaire, only 5 per cent of the youth believe that friends are more important than their family, while 25 per cent believe that family is more important than friends. Even more interestingly, the remaining 70 per cent believe that friends and family are equally important. I saw this wish – to live a life that allows for both friendships and family relations – expressed in many ways, and in the following I will argue that this wish is also expressed through ambiguous food practices, through which youth are trying to negotiate the contrast between friends and family.

In my ethnographic material, friendships across caste barriers are clearly the most common reason why young people from vegetarian families in Bengaluru start eating meat. Many students told me how they share their lunches with each other in the university canteens. I personally experienced how especially my Dalit informants asked me, although they knew I was a vegetarian, to eat meat together with them. Staples experienced the same among his Christian informants in Chennai: he was again and again offered meat by the same persons, although he had several times explained that he was a vegetarian (Staples 2008: 41). For many Dalits, meat is a symbol of democracy and the middle class and carries with it a feeling of equality (Chigateri 2008: 413–414). In short, sharing food, whether this food is veg or non-veg, is a way of practising an egalitarian relationship, where friendship is emphasized rather than the family relationship, into which the practice of following one's caste's food practice is embedded.

A widespread phenomenon today is meat-eating youth from lower castes who are putting social pressure on, sometimes even cheating or bullying, their vegetarian friends into eating meat. Twenty-three-year old Jharna comes from a vegetarian family of the Vaishya caste, and she associates non-veg with peer pressure during her life at college. Both her father and her brother were vegetarians during childhood, but after they entered college, they started eating non-veg. The same happened to her:

> I am just pure vegetarian from birth, and I changed to non-vegetarian only because of my friends. We used to go out after college. Most of them were non-vegetarians, like Muslims, Vokkaligas, a mixture of it. And I think it was only me and another girl, who was a Brahmin, we both were the only veg eaters. The rest of them were non-vegetarians. So the majority also plays a role you know, so they were like forcing us. 'You eat, you taste, it's really nice, just a spoon, after that don't eat' and all that, you know. Like that, I started eating.

The experience of Jharna illustrates that friendship for some youth may be felt to be at least as important as the caste background of their families. But often the changed food practice is followed by a certain ambivalence, which the following story about Suraj illustrates.

Suraj is nineteen years old and was raised as a vegetarian in a partly vegetarian family. His father and his two brothers eat non-veg, while the rest of the family, including all the women and all grandparents, only eat veg. While on a trip to Bengaluru's information technology city rival, Hyderabad, three friends brought him to an eatery which specialized in serving chicken, and Suraj told me, he wondered about what kind of dish he was served. What happened afterwards, Suraj told me in two different ways. The first version is that his friend Sarvesh tempted him into trying a bite, without Suraj knowing what he was eating. The other version, which I was told once, is that he was informed that this was chicken, after which he ate it voluntarily. Suraj had not previously had a desire to try to eat non-veg, but since then he has once in a while eaten chicken and fish, mostly on Sundays when they serve it at his college hostel. In a small food ranking exercise I made him complete, he placed non-veg quite high in his food hierarchy. But he was also very fond of vegetarian food and chose a purely vegetarian restaurant both the times we went out to eat together. Most Indian restaurants clearly advertise whether they serve veg, non-veg or both. I am therefore inclined to believe that one of the versions of the story about his trip to Hyderabad was consciously told in a way that made Suraj's shift to eating meat appear more innocent; that he did not try to eat it actively, but that it was his friend who cheated him into doing it. Both Suraj and Sarvesh emphasized to me how it is natural that friends share food with each other. In a different context, my informant Thakur noted that many young people do not tell their parents about their experiments with eating meat, and even though parents may be aware of it, everyone in the family will try not to speak about it. I therefore believe that Suraj is trying to negotiate the contrast between his friendships with friends like Sarvesh, on the one hand, and the food practice of his family on the other.

A similar negotiation and mediation of contrasts took place when I went out to eat lunch with Ranjit and some of his friends at a tandoori restaurant specializing in chicken. We talked a bit about caste and religion, and Ranjit stated that, 'We believe in friendship.' Later he added that he and his friends always share all meals when going out. But he also pointed out that for health reasons one should only eat limited amounts of non-veg. Although Ranjit explicitly stated he believes in friendship instead of caste, I think his opinions on health (however scientifically correct they may be) should be seen in a family and caste perspective. His mother and grandmother only rarely eat non-veg, and both his father and grandfather, the two elder male members of his family, only eat vegetarian food. Furthermore, it should be noted that Ranjit's caste (Lombani) originally came generations back from the northern part of the state of Rajasthan, where many people are vegetarians, and where people in general eat many vegetables. According to Ranjit's father, one becomes lazy and dull by eating non-veg, and I sensed a reminiscence of this health philosophy when Ranjit, on a different occasion, said to me that it is 'good' that I don't eat meat, and that one should not eat things because of their taste; one should eat 'to live': 'If you think for your taste, then you are waste. So I never think of taste, I think of health.'

Ranjit thus tries to find a balance between eating 'unhealthy' food with his friends and being loyal towards the health philosophy that he may have internalized as something very personal, but which is still clearly inherited from his father and grandfather and therefore, to some extent, has a caste origin. When young urban Indians like Jharna, Suraj and Ranjit negotiate the veg–non-veg divide, the negotiation therefore reflects a divide between, on the one hand, friendships and often taste experiences; on the other, ideals about vegetarianism and family relations.

The ambiguity of eggs

The categories veg and non-veg are commonly recognized and used in India. It is the older generation which has the authority to define such categories and concepts, my informant Vijay explained to me:

> When it comes to chicken, all people say the chicken is non-veg. From my childhood they will tell me, that it is non-veg. So I have to decide that this is non-veg only. If I tell this is veg, no one will listen. Whatever is going on here in India, whatever they tell me, when I am very small, I have to listen to that only.

Among the current generations of parents and their parents, there is a widespread consensus that eggs are *non-veg*. One young woman, Kimaya, thus told me that her grandmother had explained to her that there is life inherent in eggs, and that an egg therefore should be considered as an unborn chicken and therefore non-veg. Jharna told me that her mother was of the same opinion: 'My mother doesn't eat eggs, and even now she has that thing, like, "egg is non-vegetarian", things like that. She is very orthodox, and she doesn't like it.'

All companies in the Indian food industry as well as many food chains and other shops selling food use an official labeling system, where a green dot indicates that a product is 100 per cent veg, that is, without eggs. A red dot signifies that the product contains meat, fish or eggs. When Pizza Hut arrived in India, the cakes on their menu were all baked with eggs, but some years later they changed the recipe, which helped them sell more cakes, a manager of a Pizza Hut branch explained to me. Likewise, the ice cream chain Baskin Robbins has a big green dot on the façade of all their outlets. Interestingly, when I asked my informant Deepa, a vegetarian Brahmin, about this labeling system, she did not know what it stood for; she had just always been convinced that it was a sort of 'quality stamp'. Seen with the eyes of a vegetarian Brahmin, it is indeed a sort of quality stamp, but it is interesting that she did not know the specific meaning. It might have something to do with the fact that she only recently left her parents' home.

Among Indian vegetarians, egg consumption has in the past not been accepted as a 'vegetarian' practice.[2] But during my fieldwork it became more and more evident that eggs for many young Hindus – also Brahmins – are becoming detached from the non-veg category. My questionnaire documents this change: 44 per cent of all self-defined vegetarians consume eggs. Even among vegetarian Brahmins, a significant minority of 30 per cent eat eggs. This change appeared almost radical when I asked the respondents how they would categorize eggs. Even the answers of the Hindus pointed in all directions: 39 per cent answered non-veg, 35 per cent answered veg, 18 per cent answered 'I am confused', and 8 per cent 'Don't know'. In other words, 61 per cent of the young Hindus in my survey did not unambiguously recognize the traditional classification system in which eggs are perceived as non-veg.

Several of my informants expressed their doubts about eggs when I interviewed them, including Vijay:

> Personal opinion it is non-veg. This is non-veg only, because it comes from the chicken, so it is considered as a non-veg. ... My friend used to ask me the same question. Sometimes I will have the egg, I will say: 'This is vegetarian.' Sometimes he will tell me: 'Want to have the egg? You will call vegetarian for egg. But from egg only will come chicken, so that you will call non-veg.' He will ask so many questions. Sometimes some people think he is a stupid person, but it is a very interesting question. Because egg isn't vegetarian. From this only the chicken will come out, so we will call non-veg. How is it possible? I don't know the answer.

Suraj started stating that eggs are non-veg, but later he changed his explanation and said that an egg is something in between veg and non-veg. And Mehul pointed out, the doubt about eggs should have consequences for the classification of milk and honey:

> Brahmins will not eat non-veg, but they will drink milk. As an Indian, our culture has just commonly thought it is veg. It is not my opinion. ... Scientifically, this is a product of an animal, and this is a product of bees. I think it is non-veg.

In a similar vein, Vijay noted that milk comes from 'cow tits' and 'blood' and therefore is an ambiguous substance:

> Milk is vegetarian, but it comes ... from cow tits, from blood, it also comes from the animal, so why it is vegetarian? Why it is not a non-veg?

I think these comments on milk and honey should be seen as a way for lower-caste youth like Mehul and Vijay to express their opinion that upper-caste vegetarians are hypocrites. However, the doubts regarding eggs are obvious. My survey data clearly shows that even upper castes disagree among themselves regarding the status of eggs. Among 'pure' vegetarians, 49 per cent believe that eggs are non-veg, while only 19 per cent acknowledge eggs as veg. But among eggetarians, those who do eat eggs, only 19 per cent perceive eggs as non-veg, while as many as 47 per cent see eggs as veg. This suggests that the reclassification of eggs to a large degree is made by those who perceive themselves as vegetarians but who want to eat eggs.

Originally, I became aware of the question of eggs when I made a smaller survey at the beginning of my fieldwork and had not considered the special status of eggs. My informant Amit, in the absence of a separate egg category, had checked the box for those who eat chicken, although he only eats 'eggetarian' food. Amit grew up in a completely vegetarian family, but one day his friends asked him if he would not try to eat eggs. He asked his parents, who consented, and on the same afternoon he tasted his first egg. He was fourteen years old at that time. Another egg was eaten the same evening, and after that he ate eggs daily for an extended period of time. Today he eats eggs a few times a week. But he has never tried to eat any other kinds of non-veg, and his friends has respected this and never tried to pressure him into it. For Amit, the line is thus drawn between eggs and meat; a practice which his friends can respect and his parents accept. The ambiguity of eggs can, in other words, contain both friends and family. Amit himself noted how veg foods have 'moved closer' to non-veg foods, and that he is the first in his family to eat anything else but *pure veg*. He added that chicken does look tasty and smell delicious, and that he, depending on his mood, perhaps will start eating meat someday. However, Amit always chose *pure veg* restaurants when we ate out together, and during a food festival at his college he was disappointed about the lack of veg dishes. I therefore believe his statements should be interpreted rather as a symbolic openness than an actual wish to eat meat.

It is indeed my impression that many 'pure' vegetarians start eating eggs without later crossing further boundaries. There were no cases where eating eggs was the first step towards eating other forms of non-veg. More often, it is rather a strategy of negotiation and mediation that makes it possible for middle-class youth to eat cakes made with eggs at coffee chains such as Café Coffee Day. Officially, Deepa does not eat eggs, and when I told her to make a hierarchy of different food ingredients, she placed eggs at the bottom of the hierarchy together with other forms of non-veg. But when I then asked her if she would eat cakes made with eggs, she moved eggs a bit higher up in the hierarchy. In that very moment, we sat at a Café Coffee Day, sharing a brownie made with eggs. She never directly said that she was eating eggs but the episode illustrates that with regard to eggs, she does as many other youths who apparently do not want to miss the opportunity to enjoy chocolate cake at a modern café, when one is there anyway having a smoothie or café latte.

Changing the status of eggs in the food hierarchy gives Deepa the possibility of going out with her friends and tasting something new without feeling that she trespasses the boundaries of her family background and religious ethics. Among the parents of the young generation, this attempt at redefining categories is not very widespread. For instance, I experienced how Deepa's mother asked me whether I eat eggs. With her tone she indicated, that I, according to her perception of the food categories, was not a vegetarian.

The egg consumption of young Indians is not only interesting as a practice that challenges perceptions of food categories; it is especially interesting because, as I see it, it expresses a deeper symbolic challenge towards the whole idea of a veg–non-veg dualism and ultimately purity-based casteism. One of my informants, Ranjit, made a quite explicit link between egg consumption and the end of casteism, when he said the following: 'They used to prefer casteism, that this is non-veg, that we should not have all those things. But now people are having eggs also.' I am not sure his utterance was meant in a literal sense.

But it was, arguably, his way of saying that changing food habits are challenging the casteism of caste society.

Lévi-Strauss uses the concept *intellectual bricolage* about a form of creative composition of elements from a heterogenous repertoire (Lévi-Strauss 1969: 27). Using this concept, Mary Bucholtz has noted how youth often through bricolage create a semiotic conversion with new combinations of existing meanings connected to objects of the dominant culture (Bucholtz 2002: 536–537). Bricolage is thus, according to Bucholtz, a syncretistic practice based upon overlapping sources and contexts, where the youth reshape their own cultural background and thereby create new youth styles (Bucholtz 2002: 542). I argue that the youth in Bengaluru are, with their ambiguous egg practices, engaged in such a semiotic conversion, a mediation destabilizing the veg–non-veg dualism of caste society, and opening up the possibility of living in new ways.

Casteism or personal well-being?

It was a late evening in Bengaluru. Two of my young informants, Vijay and Deepa, had once again invited me to eat delicious kachori snacks at a small street food stand, and we were walking home through Malleshwaram, an area of the city traditionally dominated by Brahmins. I invited them for an ice cream, and the ice-cream vendor told me that his name was Khan. Jokingly, I told him: 'Oh, Khan as the famous Bollywood actor, Shah Rukh Khan.' However, the vendor replied: 'No, Shah Rukh Khan is a Muslim. I only eat vegetarian food.' The message was clear: Muslims eat meat, a true Hindu does not. Afterwards, Vijay was upset about what the vendor had said. 'Who is he to interfere in this?' And he expressed his frustration over 'such people'.

Vijay himself comes from a meat-eating family, belonging to the oil-presser (Ganigaru) caste. However, he chose to become vegetarian out of love for his girlfriend, Deepa, who is a Brahmin coming from a vegetarian family. That decision has not been well received by Vijay's family members, who scold him, arguing that non-veg gives strength and that he will become weak if he does not eat non-veg. They therefore use any given opportunity to try to convince him to start eating non-veg again. Vijay told me that he does believe that non-veg is good and healthy, and that he himself is in good health – he is rarely sick and he can tolerate low temperatures – because he grew up on non-veg food. He further commented that all members of Deepa's (vegetarian) family are very skinny and often sick. Nonetheless, although Vijay agrees with his family on the good qualities of non-veg, and although he despises the casteist ideology of the ice-cream vendor, he insists on eating veg only – even if this complicates his relationship with his family members:

> Everybody used to eat non-veg. Even they are insisting me to have non-veg also, but I don't like to have it. … I don't like my relatives. I never used to like them. When I was in my childhood, maybe when I was

very young, I used to like them very well. But when I matured, I came to know what is right, what is wrong, so I don't like them. Sometimes they will prepare non-veg, and they will call me: 'Come and have the dinner here only.' But I won't go, because I have decided to not have it.

What is clear from Vijay's words is that although he is upset about the caste ideology of some vegetarians, he is also upset about his own family's non-veg ideology. When I asked if he would someday again consider eating non-veg, he answered that this would be all up to Deepa: 'If she agrees, or if she turns, I can do that. Until then, I won't go to have non-veg.'

In short, Vijay is negotiating the essentialist food ideology of both upper castes and lower castes. Such negotiations take place at micro level every day. When I met with four young women from middle-class families from the Vokkaliga caste, which traditionally is a non-veg-eating caste, I met with them on one of those weekdays where they, according to their beliefs, are allowed to eat non-veg. Yet they chose to eat at the restaurant Adigas, which is 100 per cent vegetarian. They explained that they generally preferred to eat vegetarian food because of the taste, and that they would normally only eat non-veg on Sundays, and in that case only at home and prepared by their mother.

One way of analyzing the above-mentioned examples could be to postulate that these are simply cases of sanskritization, whereby some castes change behaviour hoping to qualify themselves for a position higher up in the social hierarchy (see for example Srinivas 1952). However, such an analysis, which reinforces the idea of a Brahmin food hierarchy ideology permeating Indian society, fails to grasp all the other possible factors that may motivate people to do as they do. When Vijay says that he has become vegetarian out of love for Deepa, we should not necessarily suspect that there is some sanskritization aspiration hidden behind this. Rather, as I argued above, vegetarianism might be a way in which Vijay distances himself from his family and from *any* caste affiliation; a distancing from caste society itself.

Jakob Klein notes that it is problematic to consider vegetarianism in South Asia as only rule-based, because 'the adoption of a meat-free diet might be related to a person's strongly-held convictions or aspirations for personal fulfilment and well-being', not only in the Global North, but *also* in South Asia (Klein 2008: 200).

Conversely, Chigateri notes how the insistence of some Dalits that beef is an important part of Dalit culture to be celebrated is as problematic as the purity ideology of vegetarians because such a way of thinking categorizes vegetarian Dalits as inauthentic Dalits (Chigateri 2008: 28). While one of my Dalit informants Mehul, who was politically active, would happily eat beef, another Dalit informant, twenty-three-year-old Hemant, would not even consider it:

RC: But do you want to eat beef?
Hemant: No no no no. I want to eat, I always use, chicken, fish, that's all.
RC: No beef...?

Hemant: No. I don't like beef. I don't like pork. Because my family doesn't like to eat.
RC: Did you try it?
Hemant: No no no no. I don't like to try.

This rejection of even the thought of eating beef sounds more like a Brahmin than a Dalit. It was confirmed by a small exercise where I asked Hemant to rank foods. While Mehul put beef at the very top of his food hierarchy, Hemant put beef at the very bottom, together with pork. Hemant furthermore argued that non-veg should be eaten with moderation, and that vegetarian food is good for health.

It should of course be noted that in cities people can choose among different kinds of food, while, as Hemant drily noted, 'In the villages, they eat food.' Seen in this way, Dalits may be eating beef due to poverty and not because they want to (Chigateri 2008: 24). (In India, beef is cheaper than other kinds of meat and sometimes comparable in price with, for example, pulses.) Interestingly, this lack of choice is apparent at student hostels where most Dalits are staying, although with the reverse effect: they are served non-veg only once a week. In short, for poor Dalits, meat consumption may not be a matter of identity or politics; it is rather a question of what is *available*, whether that is veg or non-veg.

I once asked Mehul, my highly politically engaged Dalit informant, whether vegetarians by definition are casteists. He replied:

Commonly, if I am vegetarian, I will not be casteist. But Indian culture, basically, from the origin, from the birth of the person, the caste mindset has pushed and influenced him. ... Everybody in India thinks at least fifty times per day as a caste.

Between the lines of this answer lies, as I interpret it, Mehul's recognition that essentialist ideologies about caste and food practices exist at all levels of caste society – among vegetarians as well as meat eaters. What I would like to stress, however, is that such food-based casteism is also challenged and negotiated – by both vegetarians and meat eaters.

This insight may have important consequences for the outlook of future meat consumption in India. Because if vegetarianism is generally perceived as a suppressing, casteist ideology, then as caste structures gradually weaken over time meat consumption could rise significantly. However, according to my data from Bengaluru, the food practices of youth are rather characterized by the negotiation and mediation of the contrast between vegetarianism and meat consumption. In this setting, there seems to be space for both largely vegetarian practices and some (limited) rises in meat consumption.

Conclusion: the sustainability of the (partly) holy cow

In this article, I have shown how the youth from among the middle classes in Bengaluru, in creative ways, try to unite friendships, romantic relationships, health beliefs, purity beliefs and family relations through food practices. I argue that these practices could be seen as a sort of bricolage – driven by creativity and hope; hope in the sense described by Morten Axel Pedersen: 'a prospective momentum' with the ability to connect otherwise dispersed substances, movements and networks in fragile assemblages (Pedersen 2012: 14). The ambiguity of eggs, selective vegetarianism and contextual meat consumption are thus hopeful practices through which the contrasts of veg and non-veg, and of caste society in general, are negotiated and mediated.

The practices surrounding veg and non-veg have a vast impact on the use of food resources, and by implication, environmental sustainability and human nutrition, both within India and globally. In many parts of India, a common greeting is 'Did you eat?' reflecting the historical and current prevalence of undernutrition. In fact, India still has 214 million chronically undernourished citizens (FAO 2013), and the country is depending on large amounts of food imports. This means that changes in meat consumption patterns in India are likely to have serious consequences both domestically and globally.

Marvin Harris famously argued that the holiness of the cow is a social adaptation to an ecological requirement of not eating too much meat (Harris 1978). Harris explained how cows are more ecologically and socially useful alive than dead, for example through providing milk, fertilizer and dung, while also giving birth to oxen that are used to pull the plough in the fields. He went on to conclude: 'Practices and beliefs can be rational or irrational, but a society that fails to adapt to its environment is doomed to extinction. Only those societies that draw the necessities of their surroundings without destroying their surroundings, inherit the Earth' (Harris 1978: 210).

In what has later been known within academia as 'the controversy over the holy cow', Frederick Simoons heavily criticized Harris' analysis for misrepresenting the ecological realities. Crucially, Simoons (1979) pointed out that a substantial amount of edible meat was not being eaten and therefore wasted, and that such waste of edible meat undermines Harris' view of cow holiness as something sustainable. After dismissing the ecological basis of Harris' analysis, Simoons instead argued that, to the extent cows are holy in India, it is due to particular political, religious and historical circumstances.

However, while I agree that Harris oversimplified his analysis, in a sort of protein materialism for which he has deservedly been criticized (see for example Diener et al. 1980), his insistence that Indian food practices have ecological effects, together with some social effects that go beyond caste, merits serious attention. Even the Indian Constitution gives a clear hint in that direction: the slaughtering of cattle may be allowed depending on their 'use value' (Chigateri 2008: 15). While this concept of 'use value' is open to interpretation and political conflict (Robbins 1999), it shows that even in the Indian Constitution an idea of a sustainable use of resources does somehow play a role in the perception of cows.

What makes the institution of the holy cow largely sustainable is exactly that it is *not* entirely holy at all times for all people. And since most cattle in India are still mostly fed on human-inedible by-products and not reared on resource-intensive feed, tendencies in India towards more liberal beef consumption should not be an environmental concern. This, as long as the cattle are still being produced in a manner that the FAO defines as 'default livestock', that is 'animal products and services that arise as the integral co-product of a wider agricultural system' (Fairlie 2010: 43). Neither do I believe that the concern in the near future is mutton, which is still, most likely due to price constraints, only consumed to a very limited extent, nor pork, which is not produced industrially and only consumed among some Dalits, Christians and Scheduled Tribes. The main concern regarding the sustainability of meat consumption in India is rather the consumption of chicken, which is now being produced on an industrial scale and fed with feed resources that either could have fed humans or have been grown on soil, which could instead have been used to produce food for humans. Chicken production in India has doubled between 2006 and 2015 (Index Mundi 2015).

One might argue that in comparison with meat consumption levels anywhere else in the world, and as discussed in Chapter 1, it is out of proportion to look critically at rising meat consumption levels in India. However, a sustainable global average level of meat consumption, which is only around 12 kg of meat per capita a year (Fairlie 2010: 39), is only slightly higher than current levels in India. And India does have a very large population, which makes the aggregate Indian demand for meat very high.

As this chapter has demonstrated, it is time we pay very close attention to the patterns and practices of vegetarianism and meat consumption in India. In doing so, we have to understand the challenge of apprehending how the complexity of factors such as caste, family, friendship, class, love, gender and no doubt also urbanization and geography are negotiated in everyday life by individuals, thereby creating a vast diversity of food practices.

Acknowledgements

I am indebted to the following people for providing valuable and critical input to my analysis: Marlyne Sahakian, Maansi Parpiani, Morten Axel Pedersen and Marianne Søndergaard Winther. I am also thankful towards A.R. Vasavi and the National Institute of Advanced Studies in Bengaluru, who facilitated my access to the field.

Notes

1 Parts of the material presented here have previously appeared in my Master's thesis, see Dragsdahl, R.-C. 2012. *Kød, kosmos og kærlighed i Bangalore – en antropologisk*

analyse af hvordan unge, urbane indere forhandler kastesamfundets kontraster (Meat, marriage and meaning – an anthropological analysis of how urban Indian youth negotiate the contrasts of caste society). Master's thesis, Master of Science in Anthropology, Department of Anthropology, Faculty of Social Sciences, University of Copenhagen. Defended on 16 April 2012.

2 The Lingayath caste (jati) has a reputation for being vegetarians (Chigateri 2008: 20); which, according to several of my informants, do eat eggs. They are therefore not recognized as real vegetarians.

References

Appadurai, A. 2006. 'Is Homo Hierarchicus?' In *Caste, Hierarchy and Individualism. Indian Critiques of Louis Dumont's Contributions*, edited by R. Khare, pp. 177–191. Delhi: Oxford University Press.

Bucholtz, M. 2002. 'Youth and cultural practice.' *Annual Review of Anthropology* 31: 525–552.

Caplan, P. 2008. 'Crossing the veg/nonveg divide: Commensality and sociality among the middle classes in Madras/Chennai.' *South Asia: Journal of South Asian Studies* 31(1): 118–142.

Chigateri, S. 2008. '"Glory to the cow": Cultural difference and social justice in the food hierarchy in India.' *South Asia: Journal of South Asian Studies* 31(1): 10–35.

Das, V. 2006. 'The anthropological discourse on India: Reason and its other.' In *Caste, Hierarchy and Individualism. Indian Critiques of Louis Dumont's Contributions*, edited by R. Khare, pp. 192–205. Delhi: Oxford University Press.

Desai, A. 2008. 'Subaltern vegetarianism: Witchcraft, embodiment and sociality in Central India.' *South Asia: Journal of South Asian Studies* 31(1): 96–117.

Diener, P., K. Moore and R. Mutaw. 1980. 'Meat, markets and mechanical materialism: The great protein fiasco in anthropology.' *Dialectical Anthropology* 5: 171–192.

Dolphijn, R. 2006. 'Capitalism on a plate: The politics of meat eating in Bangalore, India.' *Gastronomica* Summer 2006: 52–59.

Donner, H. 2008. 'New vegetarianism: Food, gender and neo-liberal regimes in Bengali middle-class families.' *South Asia: Journal of South Asian Studies* 31(1): 143–169.

Dumont, L. 1966. *Homo Hierarchicus: The Caste System and Its Implications*. Chicago: Chicago University Press.

Fairlie, S. 2010. 'Default livestock.' In *Meat: A Benign Extravagance*, pp. 35–43. White River Junction: Chelsea Green Publishing.

Fiddes, N. 1991. *Meat: A Natural Symbol*. London: Taylor & Francis.

Food and Agriculture Organization. 2009. *The State of Food and Agriculture 2009: Livestock in the Balance*. Rome: FAO.

Food and Agriculture Organization. 2013. *The State of Food Insecurity in the World 2013. The multiple Dimensions of Food Security*. Rome: FAO.

Harris, M. 1978. 'India's sacred cow.' *Human Nature* 1(2): 28–36.

Index Mundi. 2015. *India Broiler Meat (Poultry) Production Per Year*. www.indexmundi.com/agriculture/?country=in&commodity=broiler-meat&graph=production (accessed 5 June 2015).

Khare, R. 1992. *The Eternal Food: Gastronomic Ideas and Experiences of Hindus and Buddhists*. Delhi: Sri Satguru Publications.

Klein, J. 2008. 'Afterword: Comparing vegetarianisms.' *South Asia: Journal of South Asian Studies* 31(1): 199–212.

Lévi-Strauss, C. 1969. *Den Vilde Tanke*. Translated by Hans Peter Lund. Copenhagen: Gyldendal.

Michelutti, L. 2008. '"We are Kshatriyas but we behave like Vaishyas": Diet and muscular politics among a community of Yadavs in North India.' *South Asia: Journal of South Asian Studies* 31(1): 76–95.

Nisbett, N. 2007. 'Friendship, consumption, morality: Practising identity, negotiating hierarchy in middle-class Bangalore.' *Journal of the Royal Anthropological Institute* 13: 935–950.

Osella, C. 2008. 'Introduction.' *South Asia: Journal of South Asian Studies* 31(1): 1–9.

Peace, A. 2008. 'Meat in the genes.' *Anthropology Today* 24(3): 5–10.

Pedersen, M. 2012. 'A Day in the Cadillac: The work of hope in urban Mongolia.' *Social Analysis* 56(2): 136–151.

Raghavan, S. 2015. 'India on top in exporting beef.' *The Hindu*, August 10. www.thehindu.com/news/national/india-on-top-in-exporting-beef/article7519487.ece (accessed 7 June 2016).

Rifkin, J. 1992. *Beyond Beef: The Rise and Fall of Cattle Culture*. New York: Penguin Books.

Robbins, P. 1999. 'Meat matters: Cultural politics along the commodity chain in India.' *Cultural Geographies* 1999(6): 399–423.

Rowden, R. 2011. *India's Role in the New Global Farmland Grab. An Examination of the Role of the Indian Government and Indian Companies Engaged in Overseas Agricultural Land Acquisitions in Developing Countries*. New Delhi: GRAIN and the Economics Research Foundation.

Simoons, F. 1979. 'Questions in the sacred-cow controversy.' *Current Anthropology* 20(3): 467–493.

Srinivas, M. 1952. *Social Change in Modern India*. Cambridge: Cambridge University Press.

Staples, J. 2008. '"Go on, just try some!": Meat and meaning-making among South Indian Christians.' *South Asia: Journal of South Asian Studies* 31(1): 36–55.

Twigg, J. 1984 [1983]. 'Vegetarianism and the meanings of meat.' In *The Sociology of Food and Eating: Essays on the Social Significance of Food*, edited by A. Murcott, pp. 18–30. Aldershot: Gower.

United Nations Population Division. 2012. *World Population Prospects: The 2013 Revision*. http://esa.un.org/unpd/wpp (accessed 3 May 2014).

Yadav, Y. and S. Kumar. 2006. 'The food habits of a nation.' *The Hindu*, August 14. www.thehindu.com/todays-paper/the-food-habits-of-a-nation/article3089973.ece (accessed 5 June 2015).

Part III
Food waste dynamics

8 Uneaten food

Emerging social practices around food waste in Greater Tokyo

Atsushi Watabe, Chen Liu and Magnus Bengtsson

In 2013, *washoku* (Japanese dietary practices) was added to UNESCO's Intangible Cultural Heritage list. Due partly to some misleading news reports, this was generally misunderstood as referring to iconic Japanese dishes, such as *sushi* and *tempura*. However, what was registered was 'not a specific cuisine or dish', but the 'form of our daily home meals (consisting of rice, soup, side dishes and pickles), dietary customs for annual events, festivals and ceremonial occasions that strengthen bonds between people in local communities, and local specialty dishes' (Kumakura 2014). One of the pictures of *washoku* on UNESCO's website thus presents the scene of a family dinner with dishes of rice, miso soup and grilled salmon, and a few side dishes of boiled vegetables and pickles. The whole family – father, mother and two children – is sitting at a table, putting their hands together presumably saying *Itadakimasu*, the Japanese word used before eating to express gratitude for the gift of nature and to those who produced and prepared the food. The other pictures show scenes of a community gathering to prepare rice cake to celebrate the New Year, or the distribution of school lunches where a bottle of milk is provided to each child together with rice, soup and a main dish (UNESCO 2013). These pictures tell us that *washoku* comprises various social practices related to preparing food and having meals, including relatively new practices. It is a mix of 'tradition' and more recent additions, including more Western, ready-cooked and imported foods. That *washoku* was proposed as a Cultural Heritage probably reflects a concern that such practices are endangered; indeed, Japanese food consumption practices – both in terms of diet and how, when and with whom people eat – have changed drastically since the mid twentieth century.

This chapter aims to contribute to the understanding of how changes in food consumption patterns, practices and systems of food provision in Japan have led to an increased generation of food waste. We suggest that ordinary consumers are distanced from the sources of waste along supply chains, and that waste-reducing techniques that were commonly used in households in the past have gradually become impractical. We show that new and innovative social practices have emerged in their stead and help reduce waste to a certain extent. The chapter is based on a literature review, official data on food consumption and food waste in Japan, as well as group interviews. The twelve interviewees are diverse in gender, age (from their twenties to eighties), place of origin (from rural areas, to central Tokyo) and occupation. They were asked about their memories of practices related to preparing, eating and wasting food, including what they learnt from their parents. Each group interview involved two to six people, and lasted between sixty and ninety minutes.

Food preparation and eating as a fluid social practice

This section describes how food consumption has evolved over time, including the Westernisation and diversification of Japanese diets, as well as an increase in eating out and purchasing prepared foods. These trends provide the background needed for understanding the issue of increasing food waste in Japan today.

Westernisation of diets and growing import

According to recent surveys among Japanese youth, bread has become more popular for breakfast than rice. The most popular food for lunch is rice accompanied by a main dish, followed by Japanese noodles, curry with rice and pasta. For dinner, sautéed vegetables or meat are most popular, followed by deep-fried foods, salads, stews and curry with rice (Norinchukin Bank 2014). Surprisingly, the most popular foods for breakfast and dinner were still rare in Japan just a few decades ago.

Red meat, eggs and fried foods were introduced into Japan in the late nineteenth century by Western influences. They were mainly found in cities before World War II but spread across the country through post-war aid, mainly from the USA. School lunches were first provided by a non-governmental organisation and later by UNESCO, providing aid in the form of bread, non-fat powdered milk, cheese, meats, beans, tomatoes, grapefruit juice and many other food items that most Japanese had never even seen before. The 1950s treaties between the Japanese and USA governments required Japan to use a portion of the financial aid to purchase surplus crops from America (Kishi 1996). Introduction of Western food served to improve the dietary balance of Japanese people, who at that time were eating excessive carbohydrates, and limited fat and protein.[1]

After the post-war aid ceased, the increase in Western food continued along with growing incomes throughout the late 1950s to the mid-1970s, due to several factors. Western dishes served beautifully on the tables of middle-class families were shown in television programmes and became symbols of modern life. The evolution in housing and home appliances also contributed to this trend: new homes often had a dining room connected to a kitchen, furnished with a built-in sink and gas stove, as well as refrigerators, electric rice cookers, and, in later years, microwave ovens. The new home design and appliances enabled housewives to prepare several kinds of hot dishes and easily deliver them to the table, to waiting husbands and children (Kishi 1996; Yamaguchi 2001). From the 1950s, an increasing number of textbooks, magazines and television programmes introduced new dishes and encouraged housewives to cook from scratch every day. Housewives took to choreographing a happy dinner for their families by preparing different menus daily, including novelties like stew, hamburg steak and spaghetti (Yamao 2004; Mogi 2001).

As Western foods gradually became more widespread, the per capita supply of traditional food such as rice, vegetables and seafood decreased, and the per capita supply of meat and dairy products increased. Cereal supply has continuously decreased, especially rice, which has decreased by almost half over the past fifty years. Meanwhile, per capita supply of meat, dairy products, eggs and fat increased 5.8, 4.0, 2.7 and 3.2 times, respectively (Figure 8.1).

In parallel, the calorie-based self-sufficiency ratio in Japan dropped from 70 per cent in 1965 to 40 per cent at the end of the twentieth century. Major crops, such as wheat, corn and soy are almost fully dependent on imports, with rice as the major exception. This implies that many traditional Japanese food products, such as *udon* (wheat noodle), *natto* (fermented soy beans) and *tofu* (bean curd) are now made from imported products.

The transformation of the economy and society during the rapid growth period of the post-war era was the main reason for the continuous increase in imports, coupled with a decrease in rice consumption and the passage of

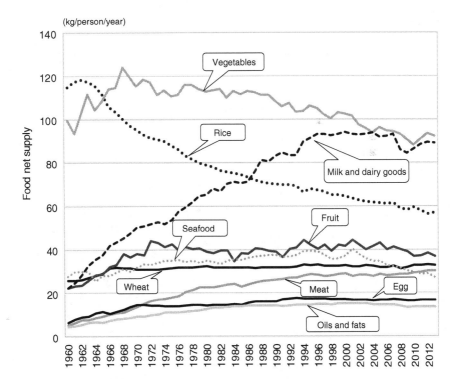

Figure 8.1 Food supply in Japan by main category, 1960–2013.
Source: Ministry of Agriculture, Forestry and Fisheries.

international trade agreements. As almost all feed grains are import-dependent, even the domestic production of meat and dairy led to a decrease in the self-sufficiency ratio[2] – coupled with decreasing rice consumption (Shogenji 2013). Changing trade conditions also meant a heavy dependency on imported foods: an increased yen exchange rate after 1985 boosted the import of fresh food, such as fruits and vegetables. The government of Japan also bowed to pressure from the United States and liberalised imports of farm products one after another, including oranges and beef. Import tariffs for most farm products have been reduced since the mid-1990s under the World Trade Organization (WTO) framework.

Increase in eating out and purchasing of cooked food

Eating out and purchasing cooked food – or the 'outsourcing of cooking' – have also increased significantly. While Engel's coefficient (the proportion of household income spent on food) declined from 60 per cent in 1948 to 30 per cent in 1977, and fell to 22.9 per cent in 2005, expenditure on eating out increased from 1980 to the mid-1990s, and spending on cooked food has continued to grow (see Figure 8.2). Since the 1970s, fast food chains and family restaurants have spread across the country. These were initially fashionable places for special occasions, such as memorial days, where a family could enjoy Western cuisine. However, Japanese people, especially those living in cities, made them part of their everyday lives in just a few decades (Kishi 1996; Kon 2013).

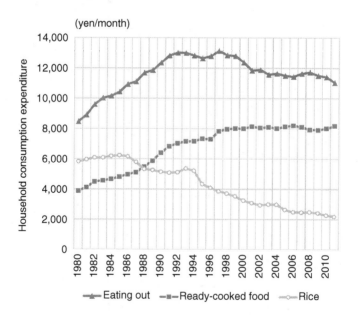

Figure 8.2 Household consumption expenditure for food in Japan, 1980–2011.
Source: Statistics Bureau, Ministry of Internal Affairs and Communications.

During the long period of economic stagnation starting in the 1990s, the market for restaurants and other eating outlets stagnated. Conversely, the demand for ready-made or half-ready meals grew steadily. Now people can find a great variety of such meals, easily accessible at any time of day or night (Murashima 2006; Beikoku Antei Kyoukyu Kakuho Shien Kikou 2014). Recent statistics show that about half of household expenditure on food is spent on eating out and purchasing cooked foods. According to a survey by the Tokyo metropolitan government in 2014, 6.2 per cent of the population dine out almost every day, and 52.9 per cent more than several days a month. Some 4.6 per cent purchase cooked foods almost every day, while 31.8 per cent do so two to three days a week. Only 4 per cent of males and 26.4 per cent of females under the age of thirty cook every day (Tokyo Metropolitan Government 2014).

The increase in 'outsource cooking' is due to various reasons, including changes in family structure. First, the average household has become smaller. Single or two-person households currently account for 60 per cent of all households in Japan. A survey shows that for single households the 'outsourcing' ratio is 51.8 per cent, largely surpassing the average of 29.3 per cent (Murashima 2006). Second, the increase in households made up of elderly people is also significant. Generally with smaller appetites, many elderly people prefer to purchase cooked foods, partly because this allows them to buy only the amount they can consume. Both smaller families and elderly households may also find it less costly to eat out or purchase ready-cooked foods than to prepare certain meals at home. Third, time spent on cooking has been reduced, partly due to the increased participation of women in the workplace. In 1970, housewives spent close to three hours cooking each day, while working women spent one hour and thirty-five minutes. In 2005, the time had been reduced to approximately two hours and one hour, respectively (ibid.). Traditional Japanese meals take time to prepare. With modern lifestyles, it is difficult to find the time, particularly in the morning rush to get to work and school.

In this section, several examples serve to illustrate changing food consumption practices in Japanese society over time, embedded in a specific cultural and social context, but also influenced by changing systems of provision. As also discussed in Chapter 11 by Keith Lee, systems of provision, coupled with the changes in the labour market and family structures, played a significant role in what types of foods were available to households.

High levels of satisfaction; new issues of concern

People in Japan today feel relatively satisfied with their 'dietary life' *(shoku-seikatsu)*. Not surprisingly, surveys demonstrate that nearly 90 per cent of respondents are satisfied with their daily meals, and about 80 per cent say that they eat what they want (MAFF 2012). Nutritional shortage and imbalances in the Japanese diet have long been alleviated, and in 1980 the Ministry of Agriculture, Forestry and Fishery (MAFF) created the term 'Japanese-style dietary life' *(nihon-gata shoku seikatsu)* – a good balance of protein, fat and carbohydrates, characterising the country's average diet (MAFF 1980). Besides having a relatively healthy balance on average, most people can enjoy a wide variety of foods, eaten at home or in restaurants.

By the 1980s, hunger had become a thing of the past and the term *houshoku jidai* (the age of plenty) appeared frequently in the media. *Houshoku* literally means getting bored with eating and is typically used in a derogatory sense to denounce the tendency to become less respectful towards food. The term is often associated with problems related to eating, such as: 1) health impacts, including increasing obesity and life-style related diseases (Matsuda 2006); 2) issues with food safety, caused by repeated incidents of contamination, poisoning and disguise of origin or use-by date (Asahi Research Centre 2007); 3) 'problematic' eating habits, such as people eating alone and family members having different dishes rather than sharing a meal together (Noguchi 2010); and 4) impacts of food production on the natural environment.

Focusing on the environmental aspect, Japan's dependency on imported foods means that the country's food miles are among the highest in the world. Vast farmland areas overseas and huge amounts of water and fertilisers are required to meet the Japanese demand (Obata 2012). A significant amount of food is wasted, along international food chains as well as in Japan.

Increasing distancing: changes in Japanese food chains

The increased distancing between farming and eating is one of the keys to understanding the sustainability issues associated with food. The notion of 'distancing' has multiple dimensions in the context of Japanese food supply. First, it refers to geographical distance, associated with high import dependence and long-distance distribution. Second, food producers are increasingly separated from consumers by multi-tiered and intransparent industrial supply chains. Because of this complexity in supply chains, consumers find it difficult to monitor threats to food safety (Shogenji 2011) as well as social and environmental issues in supply chains, including food waste. The asymmetrc information between the food industry and consumers forms psychological distancing (Otsuka 2003). Finally, many urban dwellers – especially the younger generation – have little direct experience with food production and possibly limited knowledge about how their food is generated.

Actions have been taken to address the distancing issues, both in terms of geographic distance and the loss of information across systems of provision. Direct-from-the-farm products were introduced by consumer cooperatives in 1960s in Japan, which have been recognised as forerunners in Community Supported Agriculture (CSA). Locally farmed products soon appeared also in supermarkets. In addition, over 10,000 farmers' markets exist across Japan, operated by agricultural cooperatives and other farmers' groups. Some farmers and farm enterprises also operate their own web-based enterprises and farm-direct restaurants. Local governments cooperate with farmers' groups to promote *Chisan-Chishou* (local food consumption). The *shoku-iku* (dietary education) policy introduced by the national government in 2005 also emphasises local food, and a more direct experience for consumers in relation to farming and fishing. In addition, schemes have been introduced to provide more information across supply chains, such as nutrition labelling or traceability systems, to address information asymmetry and consumer anxiety.

The growing issue of food waste

Current status and causes of food waste

The remainder of this chapter focuses on food waste, especially waste from households but starting with the broader picture. According to MAFF, the amount of food waste from the food industry in Japan was 17.6 million tons between April 2011 and March 2012 (after dehydration). Of this amount 10.46 million tons were commercially reused as fertilisers. The remaining 7.15 million tons were categorised as industrial food waste, 3 to 4 million tons of which were considered edible. The amount of food waste from households is estimated at 10.14 million tons, of which 2 to 4 million tons were estimated to be edible. Total food waste thus amounts to 17.28 million tons, about 5 to 8 million tons of which is considered edible (Figure 8.3). The edible fraction is labelled *shokuhin rosu* – edible food waste – in the statistics, a term also used here to differentiate between food waste in general and edible food waste.

It is commonly assumed that food waste increased along with economic growth.[4] However, a lack of quality data makes it difficult to trace this trend precisely. An indication is provided by the gap between calorie supply and intake, which expanded after 1970[5] (Figure 8.4). This is the period of Westernisation and diversification, and the growing habit of eating out and purchasing ready-cooked meals.[6]

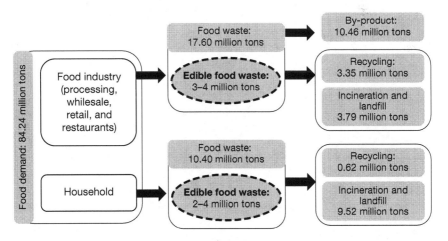

Figure 8.3 Generation and treatment of food waste (including edible food waste) in Japan, April 2011 to March 2012.

Source: Ministry of Agriculture, Forestry and Fisheries n.d. b.[3]

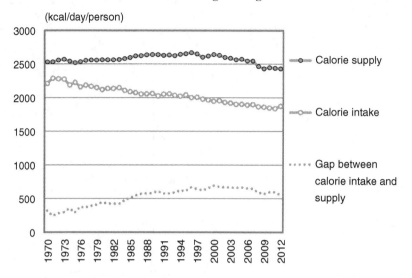

Figure 8.4 Calorie supply and actual daily individual intake, 1970–2012. The growing gap indicates an increase in food waste.

Source: Ministry of Agriculture, Forestry and Fisheries for calorie supply, and Ministry of Health, Labour and Welfare for calorie intake.

Food waste in the food industry and remedial actions

The Food Recycling Act of 2001 mandated large companies that produce, stock and sell food products to annually report the amount of food wasted. Therefore, waste generation at different stages of the food industry is monitored in detail. According to recent estimates, the food processing industry generates 16.58 million tons of waste. Wholesale generates about 222,000 tons. Retail shops waste 1.28 million tons, while restaurants generate 1.88 million tons of waste per year (MAFF 2013a; Japan Organic Resources Association (JORA) 2011). This waste can to some extent be attributed to businesses' attempts to meet consumer demands and expectations. But as mentioned above, the development of the modern food system has created increasing distancing and intransparency, leaving consumers with little information and limited direct influence.

The government and food industry have introduced policies to address food waste. Nationally, the Food Recycling Act promotes food waste recycling for fertiliser, animal feed and to generate energy. The recycling ratio in the food processing companies rose from 50 per cent in 2001 to 95 per cent in 2012, while the ratio for wholesale improved from 29 per cent to 58 per cent (MAFF 2015a; Sato 2014; Takagi 2015). The food industry has also addressed food waste through amending their business practices; they are now reviewing the rules of delivery deadlines and sales, and the best-before dates for relatively long-lasting foods (Distribution Economics Institute of Japan (DEIJ) 2014; JORA and DEIJ 2015). These initiatives certainly have impacts on food waste, however the issue of food waste from retail and restaurants largely remains.

Food waste in households

Let us now turn to waste in households. Surveys were conducted among 680 households in 2009 and 384 households in 2014 in which respondents were asked to measure and report the weight of food purchased, eaten and wasted on twenty-one occasions over one week. It was found that average households waste 41 g of edible food per person per day, or 3.2 per cent of the amount of food purchased (MAFF 2015b). This adds up to 15 kg per person annually, or 2 million tons for the total population of Japan. However, other surveys indicate that households waste a much larger proportion of food.[7] Taking these surveys into account, edible food waste is estimated to be between 2 to 4 million tons, or 2.5 to 5 per cent of total food supply (MAFF 2013b).

Leftovers in the days of plenty: deteriorating values?

Based on our interviews, people born in the 1930s or early 1940s who experienced the post-war period of hunger[8] feel that young people today have quite different attitudes towards food: they tend to order more than they can eat and leave leftovers without hesitation. This reflects the common view that Japanese people have lost their traditional instinct to avoid wasting anything of value *(Mottainai)*. In the previous section we saw the widening gap between calorie provision and intake since the 1970s, indicating an increase in uneaten food during the past decades. Also, public authorities often attribute the large amount of food waste to the changing values and attitudes brought about by increasing affluence (The Kanto Agriculture Policy Bureau, MAFF n.d.). The inter-ministry campaign to reduce edible food waste used the slogan 'Rethinking *Mottainai* at the dinner table', appealing to people's attitudes towards wastefulness.

However, is affluence the main reason behind growing amounts of food waste? MAFF's surveys tell us that the ratio of people feeling anxious about leftovers or food waste issues remains at around 80 per cent.[9] The younger generation, including those who grew up in the 1980s, have often been disciplined not to leave leftovers, and have been taught that treating food disrespectfully is a sin. A different explanation of changes in people's ways of dealing with edible food waste seems to be needed.

Skills and habits for full use of foods

Various techniques have long been used in households to make full use of foods, including husks, peels, fish bones and innards. Certain food items, such as root vegetables and smaller fish, have been appreciated because they are 'usable from end to end'. For instance, radish and turnip peels and leaves are fried or boiled for soups. Fish heads and bones can be grilled and ground, and also used for soup. Fish innards can be salted and fermented. Alternatively, all of these foods can become fertilisers, in the form of compost. While these waste-reducing practices require certain skills,[10] other practices simply involve sharing surplus food with neighbours or saving leftovers for lunch the following day.

Japanese parents have traditionally taught their children to treat foods carefully and avoid waste, and this still continues to some extent. Rice has been treated with particular respect. As the Japanese character for rice can be split up into the numeric eighty-eight, parents, grandparents, and teachers stressed the value of rice by saying that a grain of rice went through eighty-eight steps before harvesting; was brought to the table by the hands of eighty-eight people; or contains eighty-eight gods. Leaving a grain of rice in the bowl was once considered a serious offence: people recall being told that those who left rice would go blind or starve. School children were often not allowed to go out to play until they had finished their lunch and those who did not finish their school lunch were treated as sinners.[11] However, although most people have been disciplined to avoid food waste, the techniques and practices described above seem to have more or less disappeared. To understand how this has happened, we need to look at the experiences of people around family tables, which have drastically changed during the post-war period.

People born soon after the war grew up with much less diversity in terms of food compared to now. While forming their own families and raising children during the period of rapid economic growth and the bubble era, they expanded their taste and learnt about new ingredients and dishes, including various Western foods. Their efforts to produce a happy dining experience for their middle-class families were supported by the development of markets as well as the spread of home appliances. While embodying a wealthy dietary life *(shoku-seikatsu)*, they did not quickly abandon their childhood habits that made the most of food, and did not waste much. However, over time, fewer parents practised full use of food and many younger families never established such habits.

People growing up in the 1970s and 1980s experienced family dinners produced by their parents' efforts, and observed what their parents did; in most cases they were disciplined to treat food with due care and not to leave food. However, these children do not seem to have continued these practices in their own households. Interestingly, what is recalled as 'my (grand)mothers did (but I don't)' includes a wide range of practices to produce and enjoy the happy dining of the family, as well as to make the most of any foods and prevent waste.[12] Thus, the techniques to make the best use of food and to avoid waste were also part of a broader range of food-related practices, where different household members played different roles. Housewives played a key role, grandparents and husbands were usually involved to some extent, while children were usually not directly involved. We could probably better understand what makes these techniques outdated by considering changes in household composition and lifestyles.[13]

Once younger people start living away from parents, a number of factors make it difficult for them to continue what their parents did. For example, it can be too time-consuming to prepare various side dishes every day. It may also feel wasteful to do so for only one or two persons, especially since family members often have different schedules and few opportunities to eat together. So although younger generations observed and memorised their parents' food-related behaviours, it is often not feasible to continue them, since they mostly do not suit modern daily lives. Abandoning the skills and habits to make the best use of food and avoid waste can be understood similiarly. Many working women would find it embarrassing to bring their family's leftovers for lunch together with colleagues. Cooking side dishes from leftover food is still possible, but in small households even such remade dishes sometimes go bad before being eaten. And when today's urban dwellers have some leftover or excess food, they often hesitate to share it with their neighbours. Furthermore, the way of selling food has changed in line with family structure, and in response to consumer demand for convenience and time saving, allowing less opportunities to exercise these skills. Supermarkets cut off leaves from radishes and turnips, and cut and clean fish, which in the past would have been used for fish stock, for example. Vegetables and fish are packed in small portions to fit the demands of small families or single households. As discussed in Chapter 11 and in the case of Seoul, smaller portions are adapted to smaller family size. Some of these changes may have reduced household food waste, or shifted waste generation upstream to retailers and food processors.

People who grew up in the 1970s and 1980s were thus raised to treat food with due care and internalised these values, but found it challenging to continue the food waste-reducing practices of their parents.

Social responses to food waste: closing loops by connecting people

In this section, we now turn to solutions currently underway to address food waste. The movement of direct-from-the-farm products, mentioned earlier, provided opportunities for social learning about issues around food provision and consumption, including nutritions and chemicals, the decline of the agriculture sector associated with the import-dependency, as well as environmental impacts and food waste. More recent cases include the eat-all campaigns introduced by local governments in partnership with restaurants, involving consumers. Matsumoto City's '30-10 campaign', the forerunner of similar initiatives, encouraged banquet venues to request that guests remain seated and eat for thirty minutes at the start of a party, and return to their seats ten minutes before the end. This has created a social context in which people can take time to complete their servings. In addition, a number of cities have called on restaurants to offer smaller portions for the elderly and those with smaller appetites.

In addition to these initiatives replicated all over the country, some new movements have emerged and are of note. While they are not yet widespread, they demonstrate the need to address food waste in innovative ways and from a broader social perspective.

The local loop of farms and tables

Suburban areas around megacities are often home to different types of residents, including both farmers and commuter families. One such area takes advantage of this diversity to circulate kitchen garbage: Ogawa-machi, a town with many residential compounds located about an hour from central Tokyo, is known as the pioneer of organic farming in Japan. In 2001, the town introduced a scheme to recycle kitchen garbage, and turn it into biogas and liquid fertiliser. In this scheme, one hundred households segregate and dry their food waste. The city government collects and brings it to a treatment facility operated by a non-governmental organisation. The biogas from the facility is used by the residents and local farmers, with liquid fertilisers also used locally. In return for the fertiliser, the farmers issue a local currency, which residents can use to buy vegetables. Therefore, food leftovers circulate in a local loop in the form of energy and nutrients (MAFF n.d. a). Residents are directly involved in closing the physical loop, not only as consumers, but as producers of agricultural input. The local currency limits the loop to within Ogawa-machi. Participing families have direct benefit in the form of food purchased from local farms.

Food banks

Second Harvest Japan, the first and largest food bank in the country, has operated in Tokyo since 2000. Food banks collect surplus food from producers, wholesalers and shops[14] and distribute it to welfare institutions, such as orphanages and facilities for the homeless and disabled. For safety and logistical reasons, the programme does not deal with food from households (MAFF 2014). Food banks in Japan currently handle 7398 tons of food a year (MAFF 2014), a limited amount compared with the 5 to 8 million tons of edible food waste generated annually. However, the banks are still important, particularly since homeless, orphaned minors, single-parent households and other people requiring support are not very visible in society, and support institutions often struggle financially. Food bank organisations in Japan have three major reasons for operating: to address the issue of edible food waste, which they feel is *Mottainai* (regrettable to waste); to support people in difficulties; and to revitalise local communities by engaging companies and volunteers (Second Harvest Japan 2012). Although consumers cannot pass on their own leftovers, the system offers opportunities for people to support welfare facilities as volunteers delivering food. Unwanted food here works as a medium to connect people with less visible problems in local communities, while also strengthening social values around not wasting food.

Salvage parties

Events called salvage parties have been organised in Tokyo and several other places since 2013. Participants bring foodstuffs and cooked foods that would otherwise not have been used, and cook and eat together. Some of this food is seasonal gifts that the recipient does not like or know how to use. The Salvage Party Secretariat was established by young, former advertising staff, who organise parties and support other groups doing similar activities. The Secretariat sometimes brings in professional cooks, who create new recipes based on the foodstuffs available. In addition to using unwanted food that would otherwise have gone to waste, participants in the Salvage Parties share knowledge and skills, forming a basis for enjoyable meals and creating opportunities to make new friends. The Secretariat website reports on the parties, including recipes for others to use.

These three initiatives are quite different; they appeal to different groups of people and their potential for replication depends on local circumstances. But they all work in basically the same way: they put food and food waste to good use by connecting people who previously didn't know each other. They manage the physical flows by creating new social arenas where people can get together to have fun, to learn and to take meaningful action on common problems.

Discussion and conclusions

Growing amounts of food waste in post-war Japan are often attributed to increasing affluence coupled with a decreasing concern for waste. This chapter has demonstrated that the picture is far more complex, with several other factors playing significant roles. For example, the growing distance between producers and consumers – physically as well as psychologically – has made people less familiar with the causes of and the amounts of food waste generated in the long and complex supply chains of the modern food system, while also rendering them less capable of taking action to reduce waste. Furthermore, many techniques, including skills and habits, for reducing household food waste used by earlier generations have become impractical, due to lifestyle changes. Although contemporary consumers have been taught to treat food with respect and are concerned over food waste, they have difficulties acting on these values.

A number of factors have affected households' ability to reduce food waste. Time constraints seem to be an important factor with dual-earning families having less time for preparing meals and making use of peelings or leftovers. Changes in social norms have also contributed: contemporary households often hesitate to share excess food with neighbours or to bring leftovers for lunch. A growing share of ready-made food has also shifted food waste generation upstream to restaurants and food-processing facilities, out of reach of households. As a result of this 'social distance', as we have termed it, and in addition to the 'supply chain distance' mentioned above, households have fewer opportunities to apply traditional techniques for avoiding and reducing food waste.

Considering the changes described above, solutions to food waste issues cannot rely mainly on awareness-raising campaigns urging individual households to be more careful with food. Instead, solutions require a broader social perspective, and in this chapter we introduced three initiatives that seem to overcome obstacles for ordinary consumers to take action on the question of food waste: 'local loops' of farms and tables, food banks and salvage parties.

Ogawa-machi's 'local loop' connects consumers to farms, factories and shops through kitchen waste. Foodbanks create an interface for consumers to share unwanted food where it is needed. These two cases make households part of the food supply chain, albeit in different ways, and raise awareness around food waste – and ultimately its value as an important resource. Salvage parties provide opportunities to share food, with participants cooking and eating together. In this way it addresses the social distance between people, bringing people together to learn new practices. These examples show how existing attitudes towards food waste can be turned into meaningful action. They point at the need to go beyond campaigns promoting action only by individual households to also create new social interfaces. Here, unwanted food serves as a mediator between consumers, local farms and other industries (Ogawa-machi); between consumers and societal issues (food banks); or as a way to combine people's different skills and preferences (salvage parties). While the old practices of not wasting food worked well within households (sometimes involving neighbours), these new practices operate at a meso-level of society. They create new relationships between people, in producing, preparing and sharing food, as well as new stories and narratives about the value of food waste.

As an exploratory piece of work, our study did not address a number of important themes, which we will highlight here as possible areas of future research. With the ongoing polarisation of Japanese society, it seems critical to explore how food waste is generated and how it can be addressed for different social strata. From a biophysical perspective, it is important to study the composition of food waste, since the environmental impacts of food waste greatly depend on the type of food wasted. The waste generated in supply chains outside of Japan is also significant and deserves to be studied further, given Japan's high and increasing dependence on food imports. This chapter focuses on households as the entry point to food waste issues but stresses that the generation of waste from households and from supply chains are related. Changes at one stage can affect waste generation at other stages, meaning that the whole system needs to be considered. For example, a shift towards more processed and ready-made food and eating in restaurants could lead to less waste generated in households, but on the other hand the amount of upstream waste is likely to increase. There is also a need to examine the roles of restaurants, school and workplace canteens in reducing food waste at a meso-level, as discussed in Chapters 10 and 12 in this volume. From an environmental perspective, the upstream shift of waste generation might be a good thing; food-processing factories, restaurants and supermarkets are perhaps better positioned to separate useable or recyclable food, and larger volumes may enable more effective treatment. However, food waste recycling is always inferior to waste reduction, as food production is resource-intensive and recycling can only recover a fraction of those resources. This illustrates the complexity of the issue and the need to approach food waste in a systemic fashion, taking into consideration how changes at different tiers of supply chains interact and affect food waste outcomes, including both the overall amounts generated, the composition and the technical and economic potential for beneficial use. Food waste issues are not only a matter of how much is generated, and what comprises the waste, but also where in supply chains, including households, it is generated and who can control what happens to it.

Promising approaches towards reducing food waste do exist, but upscaling has limited potential, and more systemic issues need to be addressed. Researchers play an important role in highlighting the negative impacts of food waste, thereby raising awareness and helping to formulate government policies. However, research and associated policies must consider both the physical and social/cultural aspects of food consumption and related waste, recognising that food is not only made up of material flows, to be studied through quantitative analyses, but also deeply intertwined with ever-changing human practices. In this study, we have found that this includes how systems of provision develop over time, the moral dimension of wastage, the material dimension of what food is available and in what settings it is consumed, and finally the time constraints affecting everday food consumption. All of these aspects should be considered when exploring future food production and dietary habits in Japan, and how to achieve a sustainable and nutritionally well-balanced food supply for the country's population while limiting food waste.

Notes

1 The government launched the Nutrition Improvement Promotion Campaign in 1949 to address dietary imbalance involving excessive carbohydrates and insufficient protein and vitamins (Ministry of Health, Labor and Welfare 1958).
2 The calorie-based self-sufficiency ratio is equal to the daily per capita per day supply of calories that is domestically produced (947kCal in FY2014) divided by the total daily per capita per day supply of calories (2415kCal in FY2014). While animal products fed with imported feeds are counted as imported products, and stockbreeding in Japan depends mostly on imported feeds of imported cereal feeds, even the increase in Japanese stock production leads to a decrease in the calorie-based self-sufficiency ratio. Japan is one of a few countries to officially adopt this index.
3 More detailed data for the following years are accessible at MAFF n.d. c.
4 See Chapter 1 of *The Environment White Paper* 2002 (MOE 2002).
5 Per capita calorie supply slowly increased from 1970 until 1999, while the calorie intake steadily decreased. Several conditions could be assumed for the decrease in the calories consumed, including ageing population, decreasing primary and secondary sectors in the total labour force, and decreased rice consumption.
6 While these two figures are not directly comparable, due to differing data sources and calculation methods, the growing gap indicates that waste generation has most likely increased.
7 For instance, Kyoto city estimates that its citizens generate an average of 924 g of household waste a day, including 41.5 per cent of kitchen waste, based on its survey of the composition of garbage collected from sample communities (Kyoto City Government, n.d.).
8 See Chapter 9 by Dunfu Zhang in this volume for a discussion around collective social memories related to famine, and in the case of Shanghai.
9 This finding is echoed in the work of Evans (2011) in relation to London households.

10 Based on our research, other food preparation skills may also not have been trasnferred to younger generations, such as preparing several side dishes daily; preparing pickles daily; offering small portions of rice to family Buddhist altars; or serving fruit as dessert.
11 While not all families could train children to behave properly at the table, school lunches assume an important role in teaching manners, including the proper order for eating certain foods, the handling of chopsticks, bowls and dishes, or treating foods with due respect, not being picky, and not making leftovers. Teachers do not strictly force children to eat everything recently due to concerns about food allergies.
12 Practices of parents and grandparents include, for instance: introducing new recipes learned from television programmes; preparing several different side dishes every day; serving fruit as dessert; offering small portions of rice to family Buddhist altars; cooking side dishes from vegetable peelings, fish heads/bones and innards, or remaking side dishes from leftovers; pouring hot water or tea into the rice bowl after eating, scraping up the remaining rice or sauce and drinking it.
13 Hargreaves (2011), referring to Warde (2005), rightly stressed 'a need to consider particular domains of everyday life' to study 'the grounded performances and negotiation of whole bundles of practice'. Sahakian and Wilhite (2014) also pointed out the important interrelationship between practices.
14 Certain private companies, such as the large supermarket chain Seiyu, have strengthened their support to food banks through increased food donations and financial contributions.

References

Asahi Research Centre. 2007. *Shoku ha dokohe iku? Kawariyuku shoku to shoku seikatsu ni kansuru saishin jijou [How Will Our Diet Go? The Recent Trents of Our Foods and Dietary Life Under Transition]*. Tokyo: Asahi Research Centre.

Beikoku Antei Kyoukyu Kakuho Shien Kikou (The organisation to support the secure provision of rice). 2014. *Nakashoku, Gaishoku no Doukou [The Trend of Ready Cooked Meals and Eating Out]*. www.komenet.jp/pdf/chousa-rep_H26-4.pdf (accessed 29 June 2015).

Distribution Economics Institute of Japan (DEIJ). 2014. *Nouhin Kigen Minaoshi Pairotto Purujekuto Saisyu Houkoku Shiryou [The Final Report of the Pilot Project of Reviewing the Deadline of Delivery]*. Oral presentation at the final report meeting of the pilot project, 13 March 2014. Tokyo: DEIJ.

Evans, D. (2011). 'Blaming the consumer – once again: The social and material contexts of everyday food waste practices in some English households.' *Critical Public Health* 21(4): 429–440.

Hargreaves, T. 2011. 'Practice-ing behaviour change: Applying social practice theory to pro-environmental behaviour change.' *Journal of Consumer Culture* 11(1): 79–99.

Japan Organic Resources Association (JORA). 2011. *Heisei 22 Nendo Shokuhin Haikibutsu tou Hassei Yokusei Suishin Jigyou Houkokusyo [Report of the Activity to Promote the Reduction of Food Waste Fiscal Year 2010]*. Tokyo: JORA.

JORA and DEIJ. 2015. *Shokuhin rosu sakugen notameno shoukansyu kentou waakingu chi-mu ga honnendo seikawo kouhyou nouhin kigen to shoumi kigen no minaoshi susumu [The working team to review the business practices toward reducing edible food waste released the result of the Fiscal Year: continuous progress on reviewing the delivery deadlines and best before dates]*. www.dei.or.jp/aboutdei/pdf/press/press_150317_01.pdf (accessed 29 June 2015).

Kanto Agriculture Policy Bereau, Ministry of Agriculture, Forestry and Fishery. *Houshoku Nippon Tabenokoshi no Genjou, Shoku wo kangaeru* series, Vol. 4. http://www.maff.go.jp/kanto/shizuoka/panfu/pdf/vol4_1.pdf (accessed 29 June 2015).

Kishi, Y. 1996. *Shoku-to nou no sengoshi [The Postwar History of Eating and Agriculture]*. Tokyo: Nihon Keizai Shimbun Shuppansha.

Kon, T. 2013. *Famirii resutoran: gaishoku no kingendaishi [Family Restaurants: The Modern History of Eating Out]*. Tokyo: Kobunsha.

Kumakura, I. 2014. 'Washoku joins UNESCO's intangible Cultural Heritage list.' *Food Culture*, 31 March. www.kikkoman.co.jp/kiifc/foodculture/pdf_24/e_003_005.pdf (accessed 7 June 2016).

Kyoto City Government. n.d. *Kyoto shi no namagomi deeta [The Data on Garbage in Kyoto City]*. http://sukkiri-kyoto.com/gomidata/#gomidata3 (accessed 29 June 2015).

MAFF. 1980. *80 nendai no nousei no kihon houkou [The Basic Strategy of Agriculture Policy in the 1980s]*. Tokyo: The Ministry of Agriculture, Forestry and Fishery.

MAFF. 2012. *The Shoku-iku hakusho [Dietary Education White Paper]*. Tokyo: The Ministry of Agriculture, Forestry and Fishery.

MAFF. 2013a. *Heisei 23 Nendo Shokuhin Haikibutsu tou no Nenkan Sou Hasseiryou Oyobi Syokuhin Junkan Shigen no Saiseiriyou Jisshiritsu Nitsuite [Report of the Annual Amount of Waste Derived from Food and their Recycling Rate, FY2011]*. www.maff.go.jp/j/press/shokusan/kankyoi/130617.html (accessed 5 November 2015).

MAFF. 2013b. Shokuhin rosu sakugen ni mukete – Mottainai wo torimodosou – *[Toward the reduction of edible food waste – let us rethink Mottainai]*. www.maff.go.jp/j/shokusan/recycle/syoku_loss/pdf/0902shokurosu.pdf (accessed 29 June 2015).

MAFF. 2014. *Kokunai no fuudo banku no katsudou jittai haaku chousa [The survey to understand the status of activities of food bank in Japan]*. www.maff.go.jp/j/shokusan/recycle/syoku_loss/foodbank/tachiage/pdf/25fbhk.pdf (accessed 19 June 2015).

MAFF. 2015a. *Survey on the Recycling of the Circulatory Food Resources*. www.maff.go.jp/j/tokei/kouhyou/zyunkan_sigen/index.html (accessed 29 June 2015).

MAFF. 2015b. *Shokuhin rosu toukei chousa setai chousa Heisei 26 nendo* (FY2014 Survey on Edible Food Waste from Households) www.maff.go.jp/j/tokei/kouhyou/syokuhin_loss/ (accessed 29 June 2015).

MAFF. n.d. a. *Hirogaru namagomi shigenka no torikumi [The Activities to Turn Garbage into Resources Spreading]*. www.maff.go.jp/kanto/to_jyo/jyouhou/senshin/biomass/baio11-08.pdf (accessed 29 June 2015).

MAFF. n.d. b. *Shokuhin haikibutsu tou no riyou joukyou tou Heisei 23 nendo suikei gainenzu [The Conceptual Figure on the Status of Reusing the Wastes Deriving from Foods, FY2011]*. www.maff.go.jp/j/shokusan/recycle/syoku_loss/pdf/h23_flow.pdf (accessed 29 June 2015).

MAFF n.d. c. *Shokuhin resaikuru hou ni motozuku teiki houkoku no kekka nitsuite [On the periodical reports based on the Food Recycling Act]* www.maff.go.jp/j/shokusan/recycle/syokuhin/s_houkoku/kekka/gaiyou.html (accessed 9 May 2016).

Matsuda, T. 2006. 'Shoku seikatsu no henka to seikatsu shukan byou [The change of dietary life and adult diseases].' *Hokkaido Kyouiku Daigaku Asahikawa Kou Syakaigaku Kenkyushitu Chousa Houkoku [The Study Report of the Sociology Labo, Asahikawa Campus, Hokkaido University of Education]*. Asahikawa: Hokkaido University of Education Asahikawa Campus.

Ministry of the Environment (MOE). 2002. *Heisei 15 nenndo kankyo hakusho [The Environment White Paper 2002]*. www.env.go.jp/policy/hakusyo/hakusyo.php3?kid=218 (accessed 16 July 2015).

Ministry of Health, Labor and Welfare (MHLW). 1958. *Eiyou Kaizen Fukyu Undou no Ayumi [The Progress of the Nutrition Improvement Promotion Campaign]*. Eiyou Nihon (Nutrition Japan) August 1955, pp. 17–19.

Mogi, S. 2001. 'Shyouhisha niizu no henka to gaishoku sangyou no taiou [The change in the consumers' needs and the responses of the dining out industry].' In T. Hiroyuki and Y. Hiroyuki (eds), *Shokuseikatsu no henka to huudo shisutemu [The Change of Dietary Life and the Food System]*. Tokyo: Association of Agricultural and Fishery Statistics.

Murashima, K. 2006. 'Nakashoku shijou no genjou to tembou [The current status and prospects of the market of ready-cooked meals].' *Mizuho Industry Focus* 50, 28 September.

Noguchi, J. 2010. 'Gendai Nihon ni okeru shoku no kankyou to shokutaku no henka kodomo to kazoku ni shouten wo atete [The change of the environment of diet and tables in modern Japan focusing on the children and the families].' *Bulletin of the Department of Sociology* 38. Kyoto: Bukkyo University.

Norinchukin Bank. 2014. *The 2nd Survey of the Awareness of Modern Single People in their 20s on Eating Habits and Food Safety*. www.nochubank.or.jp/contribution/pdf/research_2014_02.pdf (accessed 29 June 2015).

Obata, N. 2012. 'Shoku ni kanren shita kankyou fuka no sakugen to jizokukanou na shoku to nou ni kansuru kenkyu [A study on reduction of food wastes and sustainability food system and agriculture].' *Seisaku Kagaku* 19(3): 249–267.

Otsuka, N. 2003. 'Shoku to nou no bunri niokeru senmonka to shirouto no bunri [The "professional–lay divide" in the "food–agriculture divide"].' *Journal of Environmental Sociology* 9: 37–53.

Sahakian, M. and H. Wilhite. 2014. 'Making practice theory practicable: Towards more sustainable forms of consumption.' *Journal of Consumer Culture* 14(1): 25–44.

The Salvage Party Secretariat. http://salvageparty.com/ (accessed 29 June 2015).

Sato, K. 2014. 'Shyokuhin Haikibutsu Sakugen to Shokuhin Risaikuru no Genjou to Kadai [The current status and challenges of food waste reduction and food recycling].' *Bulletin of the Department of Sociology* 48. Senshu University.

Second Harvest Japan. 2012. *Heisei 23 nendo nousan gyoson rokuji sangyouka taisaku jigyou "Fuudo banku katsudou suishin jigyou" houkokusyo [Report of the Commissioned Work to Promote Food Bank Activities]*. Tokyo: Second Harvest Japan.

Shogenji, S. 2011. 'Fuudo shisutemu ron to gendai nihon no shokuryou shokuhin mondai [The food system study and food issues in modern Japan].' *Discussion Paper Series* 4. Tokyo: Institute of Social Science, University of Tokyo.

Shogenji, S. 2013. 'Nihon no shoku to nou wo furikaeru [Looking back on the food and agriculture of Japan].' *The Hyogo Economy* 117. Hyogo Economic Research Institute.

Takagi, S. 2015. 'Shokuhin Haikibtsu tou no risaikuru ni kansuru kadai to kaiketsusaku [The challenges of food waste recycling and the solution].' *Mizuho Information and Research Institute Report* 9.

Tokyo Metropolitan Government. 2014. *Shokuiku to Shoku no Anzen ni Kansuru Seron Chousa [Opinion Survey on Eating Education and Food Safety]*. www.metro.tokyo.jp/INET/CHOUSA/2014/10/60oau104.htm (accessed 29 June 2015).

United Nations Educational, Scientific and Cultural Organization (UNESCO). 2013. 'Washoku, traditional dietary cultures of the Japanese, notably for the celebration of New Year Inscribed in 2013 (8.COM) on the Representative List of the Intangible

Cultural Heritage of Humanity.' www.unesco.org/culture/ich/RL/00869 (accessed 29 June 2015).

Warde, A. 2005. 'Consumption and theories of practice.' *Journal of Consumer Culture* 5(1): 131–153.

Yamaguchi, M. 2001. 'Shoku kukan to daidokoro dougu no henka [The change in the eating space and the kitchen appliances].' In T. Hiroyuki and Y. Hiroyuki (eds), *Shokuseikatsu no henka to huudo shisutemu [The Change of Dietary Life and the Food System]*. Tokyo: Association of Agricultural and Fishery Statistics.

Yamao, M. 2004. *Kyoumo Ryouri: Ryouri Bangumi to Shufu Kattou no Rekishi [Cooking Today as Usual: The History of Conflict Between the Cookery Shows and the Housewives]*. Tokyo: Hara Shobo.

9 From thrift to sustainability

The changing table manners of Shanghai's food leftovers

Dunfu Zhang

Chinese people, whether rural or urban, poor or rich, have historically been quite thrifty with food, mainly because of personal experiences related to hardship and starvation, and the collective social memory of famine. The planned economy of 1960 to 1980 also contributed to the development of habitual practices around saving food, such as making sure that not a single grain of rice, wheat or corn would be wasted at the table. Ever since market reform and the opening up of the economy, which has brought with it an increasing provision and variety of food products, food waste has increased during festivals, banquets and celebrations – especially those paid for by public funds or aimed towards business networking opportunities. The focus of this chapter is on the changing practices related to food consumption, leftovers and related waste in the context of macro-level socio-economic trends, including access to a global marketplace, increased participation in the labor market and rapid urbanization.

Using social practice theory as a theoretical framework and drawing mainly from informal interviews in Shanghai, this chapter argues that most people are strictly and consistently thrifty regarding how to handle food leftovers. Strong love for their family and relations of intimacy are tied to how people carry out these often speechless and stubborn habits. For some residents, managing leftovers is part of an explicit discourse: some have learned from food scarcity in the past, some manage leftovers as a way to express caring relationships with friends or relatives, while others are concerned about environmental pollution. As part of table manners, the urban Chinese, and especially professional Chinese with higher cultural capital and international experiences or perspectives, are demonstrating increased consciousness around the notion of 'sustainable consumption' – a relatively recent phenomenon. The discourse behind these changing table manners could be named 'from thrift to sustainability'.

Currently, those who practice a more thrifty approach to leftover food because of famine memory and out of family welfare are more prevalent than those who handle food leftovers with environment concern, yet these practices could change over time. In the near future, with the memory of famine fading among younger generations, the discourse around sustainability will no doubt grow. This chapter examines current perceptions and practices related to food consumption and food leftovers in the sociological perspective of sustainable consumption. By exploring and illustrating everyday routines of practices, it attempts to address the following questions: Against what social, cultural and historical background does routines around food consumption play out? What are the important features of these practices and how do they relate to national and global environmental problems? During the process of market transition and the individualization of society, what are the environmental responsibilities of Chinese people to reduce food waste? And finally, what are the challenges that China faces in its long march towards a green nation?

In the following section, I outline the research approach and conceptual framework. I then continue with research results that demonstrate how food leftovers are closely related to love and intimacy among family members and close relatives, the tradition of thrift or frugality, and relatively new environmental concerns regarding how to deal with food leftovers, and finally wedding banquets as the most important events involving food waste for common people. I then follow with concluding remarks focusing on how handling food leftovers is significant towards sustainable consumption in the global context.

Research approach

Located in the Yangtze River Delta in East China, Shanghai sits on the southern edge of the mouth of the Yangtze in the middle portion of the Chinese coast. With a population of more than 24 million as of 2014, Shanghai is among the largest and most international cities in the world, as well as a global financial center and transport hub with one of the world's busiest container ports. Shanghai is also one of the four direct-controlled municipalities of the People's Republic of China. With historical nicknames such as 'Pearl of the Orient' and 'Paris of the East', Shanghai was known as the leading 'dragon head' (*Longtou*) during China's reform era and is generally seen today as a truly global city. Shanghai was chosen as my fieldwork site for several reasons, not least its long tradition of business and trade, and consumer engagement in the consumer society, involving the acquisition and care of household decoration items, clothing and fashion, or food. The city has been considered one of the most cosmopolitan and international cuisine centers of China since the late 1980s (Farrer 2011: 36). As Lucian Pye (1981: xi) succinctly stated, 'serious analysis of nearly all important aspects of life in China must, eventually, confront Shanghai and its special place in the Chinese scheme of the things'.

Data in this chapter is based on informal interviews with thirty-six informants in Shanghai from 2013 to 2015. With so many people who are busy making a living or trying to realize their dreams of success, it is quite hard to get people talking about their food habits. On the other hand, as the leading city in business, fashion, consumption and lifestyle, Shanghai is the right place to study food. Therefore convenient sampling was applied, in most cases based on acquaintances and their relatives, friends and colleagues. In other cases, I talked with business partners, tourists, restaurant customers or managers, while working or dining. In some cases, I picked up the topic of food waste at a dinner or lunch table when dining with friends or colleagues. Focus group discussion around certain specific topics also took place. I often recorded observations in diaries, after the fieldwork and in addition to the interviews.

During my fieldwork, such questions were often asked: How do you deal with your food leftovers at home or when dining out? What goes through your mind at that time? Some observations are also included as a supplementary empirical source. These data tell me how people living in Shanghai reuse food leftovers, as well as the practice of food saving in various ways. As leftover food and its many uses is one of the least developed areas in China's sociology and anthropology studies, these phenomena intrigued me most. While most Chinese sociologists and anthropologists are focusing on grand narratives such as social transformation, globalization, state–society relationships, urbanization and social governance, I am more interested in these seemingly trivial, everyday routines. To me these everyday life practices could be important factors in a future green (or grey) planet, or a clean (or dirty) economic development for the Chinese Tiger. The issue is not only about making sure a resource such as food is maximized, but also something about how food waste is managed downstream from consumers – and how the question of increased waste is also problematic from an environmental and health perspective.

The theoretical inspirations to explore and analyze human practices could be traced back to Giddens (1984) and Bourdieu (1977). I was exposed to theories of social practice and their application to food when I was a senior research fellow at the Sustainable Consumption Institute at the University of Manchester, where these theories were discussed during seminars and group reading events. Practices are sets of 'doings and sayings'; they involve both 'practical activity and its representations' (Warde 2005: 134). In the words of Reckwitz (2002: 250): 'A practice is thus a routinized way in which bodies are moved, objects are handled, subjects are treated, things are described and the world is understood.' Social practices, then, rather than individuals, norms, cultural codes, discourses, macro-level trends and social structures become the central unit of analysis and intervention. Thus they are of more interest to sociologists, particularly for a sociology of consumption. To be specific: 'What counts is the big, and in some cases, global swing of ordinary, routinized and taken-for-granted practice' (Shove 2003: 9). Theories of practice have much to contribute to sociological analysis of sustainable consumption, placing attention on the everyday inconspicuous consumption of energy and other resources, which can have profound environmental impacts, possibly more than certain forms of conspicuous consumption. The construction of basic elements of the theories of practice was formulated in the later twentieth century, and up to now, theories of practice have made an important contribution to the study of sustainable consumption (for details, see Welch and Warde 2015). Though Chinese people have a rich variety of food production and consumption practices, practice theory is still new to China's academic community. This chapter is therefore a timely effort to apply practice theory to current-day food consumption in China.

Research findings

An old Chinese saying goes like this: 'Food is the paramount necessity of the people.' This principle has remained unchanged throughout time and has rich significance for China and the Chinese people. Feeding such a huge population is the most basic condition for nation survival, without which prosperity is out of the question. Political revolutions and military unrest also have much to do with food supply, consumption and distribution.

The workings of China are extremely complex and can be hard to understand, yet one basic fact has remained consistent over hundreds of years: the limited resources available to support such a large population. Historically, natural disasters such as floods and droughts were the principal reasons for famines, along with foreign invasions and internal rebellions, which pushed the whole country to critical points. Famines were often recorded in Chinese history, but in the late nineteenth and early twentieth centuries China experienced them with a frequency and intensity that seemed to fulfill the Malthusian prediction that overpopulation would result in catastrophes (Li 1982: 687). John L. Buck's surveys conducted in the 1920s found that peasants recalled an average of three famines during their lifetime, with hunger lasting approximately ten months, which forced a quarter of the affected population to consume bark and grasses (Buck 1937: 124–127). In the eyes of Western people, China became known as the 'Land of Famine' (Mallory 1926) at that time.

To understand China's food history, few new foods have been introduced in the past few centuries, and the basic dichotomy of *fan* and *cai*, grains and accompanying meat and vegetable dishes, has remained a constant mode of eating (Chang 1977: 7–8). As to food culture, and different from most other societies, Chinese use chopsticks rather than knives and forks, and when dining together they share food from the same table, often from the same plates or bowls. Today with more than 1.3 billion people, the sheer challenge of feeding China's population remains as urgent as it was in the past (Li 1982; Smil 1995). Although the government is proud of keeping a quarter of the world's population alive, China faces the challenge and urgency of food provisioning that no other nation in history has ever faced. In this context, how people manage leftovers from the table is a critical issue.

When finding enough food to eat was a challenge for many people, food waste did not make up the main part of Shanghai's municipal solid waste. As is the case in other cities across China, waste sorting remains a problem after ineffective governmental efforts. Yet there is increasing concern over food waste in China these past years, and for good reason – food waste is on the rise. Ever since the opening up of the Chinese market and reform policies of the late 1970s, food waste has become a headache for residents, cleaners and government officials. During summers in the 1980s, the daily amount of solid waste often exceeded 10,000 tons, because of a significant rise in food waste resulting from the availability of bulky foods such as watermelon and other fruits and vegetables (Ward and Li 1993: 31). In the early twenty-first century, over 1300 tons of kitchen waste was generated per day in Shanghai (Wang et al. 2010). Of this, more than 50 percent was not collected and transported properly by a comprehensive and formal collection and transportation system (An et al. 2014: 245). By one account, food waste comprises a staggering 70 percent of the country's garbage; by comparison, this figure is estimated at 20 percent in the United States of America.[1] Yet how these figures were generated remains unclear and no official data exists on food waste in China. This underlines the difficulty of measuring food waste quantities, as outlined in several contributions to this edited volume (see for example Chapter 10 by Papargyropoulou, and Chapter 12 by Favis and Estanislao).

Regional, cultural and personal preferences all influence the practice of food consumption and the reuse of leftover food. For example, Shanghainese make use of rice from their last meal to make vegetable rice porridge, a delicious nutritious and local cuisine that is even popular in restaurant menus. New immigrants to Shanghai from other provinces bring with them their own ways of reusing. For people from north China, such as Shandong, Henan or Hebei provinces, leftover bun slices can be fried together with eggs and converted into a brand new meal known as 'silver wrapped with gold'. For people from Jiangsu, Zhejiang, Anhui and Jiangxi provinces, yesterday's rice can become the main ingredient in fried rice with egg, or fried rice with shrimp, among other variations. Even children are fond of using the last meal's rice for another dish: some rice and tomato sauce, covered with one or two pieces of cheese, can easily be placed in the microwave to create a new dish.

Food leftovers, love and family

For certain middle-class British families and according to Cappellini (2009), food leftover consumption has analogies with the process of sacrifice. Consuming leftovers requires a high degree of admission to the family and thus only family members, and some of them more than others, are called to such a sacrifice. Therefore it is not only during extraordinary food consumption that familial bonds are reinforced, rather it is also during the everyday practice of consuming leftovers that familial bonds are sustained and perpetuated. It is quite the same with the Chinese domestic practices of consuming leftovers, be they food, clothes or other household items, but in China this is more prevalent among the lower-class peasants and factory workers and their families. Food is so important in China that it is often explicitly associated with the closest of social ties. Among older generations of women, a saying goes like this: 'Please the man's stomach before you get his heart.' In other words, as a woman, you have to prepare enough delicious meals or be an excellent cook before someone falls in love with you. Maybe that is one of the evolutionary results from centuries of household division of labor: men plow and harvest in the fields while women cook and weave at home. Sharing food (and associated leftovers) from the same table and from the same dishes is an old tradition, to such an extent that there is no surprise when somebody says, 'If a man eats a woman's

From thrift to sustainability 185

food leftovers, that means he loves her.' This happened to a young man on a dinner date:

> (I've been) in love with a girl for a long time, but I'm not ready or brave enough to say 'I love you'. Now a chance came that we go out for a meal together. She could not finish her dishes before I finished mine, while I still could take more food. I tried hard to curb my excitement, and told her: since you couldn't finish, let me help you. She proposed to order another dish. My reply was: no, that's all right for me.[2]

The young man finished his date's meal, which may seem unusual, but can happen quite often among young couples during the period of dating: once their relationship is established, the food leftovers, including the touched foods, can be packaged and brought back home by either of them.

If I recall college campus life in the 1980s, couples in love often went to the dining halls together. The amount of food provided was according to a price list. Boys, who could eat more than girls, actually got the same quantity of food. So it was common for the boys to share food from their girlfriend's bowl or lunch-box. They would sometimes feed each other from each other's chopsticks or spoons, partly to show how intimate and caring for each other they were. Their ways of food sharing are often the envy of single students. One postgraduate, Miss Cao, commented, 'It is still true now in every university dining hall, sometimes in restaurants,' that couples will share their food.

Dating often comes before food sharing, yet there are exceptions. One female office worker from a big company told me:

> I know a couple. The boy does not eat the leftovers of his girlfriend, it is the girl who saves the yummy stuff for the boy. Later they dated each other, and now they are married. It is too soon to say 'If you want to get a man's heart, save the delicious food for him in case you could not prepare it in the kitchen', but in this case, this is exactly what the girl did: she saved her food for her boyfriend to show how sincere her love was.

How to understand the ambivalence of this food sharing between or among such close connections? Mrs Dong, an education consultant in her late forties, had an explanation that sounds quite convincing:

> Many women are influenced by this psychological hint. It makes sense. For most families it is the parents, often the father, who finished the food remains. Children feel strong attachment because of this protection. For a girl of dating age, she may feel the boy cares for her if he doesn't hesitate to take her leftovers, similar to her experiences with her father.

Family members also share food leftovers. More often than not, parents finish their children's food; for older generations, it sometimes happens that the elders will take the remains of younger brothers or sisters. This type of behavior is often recognized and understood as a habit, which is very hard to change, just like Mr Hu in his mid fifties stated:

> If your parents or somebody does it, don't ask why and don't persuade them to give up. Their frugality is based on decades of habit. They finish the leftovers, yet they insist that their spouse or children will eat fresh. They even indicate that they love to eat the food remains.

This situations echoes well with the comment of a housewife in her early sixties:

> Old stories are quite popular that mom would make clear that she love fish tails, or that chicken feet is granny's favorite. By eating these parts, they can say somewhat proudly they did eat fish or chicken. Actually, they try by all means to guarantee that their husbands and children will easily have enough meat from the dish.

As a consequence of decades of this form of family love, feedback from the younger generation could be identified, for example from the following girl's wish during the several days she was away from home: 'Granny, could you please not finish the food leftovers secretly?' The girl actually feels sorry that she could not help her grandma to finish the leftovers.

Resisting being involved in food leftover sharing can even lead to the break-up of an established relationship. According to a news source, both Li and Ye were born in the 1980s; soon after they graduated from college, found jobs and were married. Ye was frugal; sometimes they prepared more than enough food and would save it for the next day, but Li insisted he would not eat 'dishes over night'. Ye realized that '(our) marriage life proved that he is so selfish'. This finally led to their divorce.[3]

When asked his response given similar situations, Mr Liang said,

> I love her and I must accept every reasonable request. Finish the dish remains is one of them. Moreover, if we have a baby, I will take care of all its food leftovers. This is my love for my wife and child. By the way, don't waste grain, or it is a shame.

There are similar stories for families. For the twenty-year old boy Liu, his teenage life sounds like a difficult negotiation with leftovers and academic performance:

> During middle school, mom forced me to eat (food leftovers). But sometimes I refused. I feel very bad about the experiences. I think it has much to do with my poor academic performance. I did not get a better score at the entrance exam and I failed to go to an ideal university.

With increasing girl power and women's rights, sufficient food supplies and stronger consciousness around hygiene, mothers and girls seems less interested in food leftovers. But in nuclear families, the father or husband's habits remain quite the same. One nine-year-old boy described his family's table pattern: 'It is daddy who takes my unfinished food, milk, bread, sausage, anything. Sometimes mom warns him not to gain weight.'

Who finishes the food remains? It depends on his or her social status within the family. One businessman in his mid-forties said,

> In my family's power hierarchy, my daughter is on the top, my wife in the middle. You know the result. I am always the last one to leave the table. Also my role in my household division of labor involves cleaning, and the easiest way of washing dishes, bowls and cleaning tables is finishing the leftovers.

Hui's family has a similar story. There is often residual milk and beans left after making soya-bean milk. On one day, over half the beans were leftover. But sixteen-year-old Hui only likes fresh beans, he would never drink the soya-bean milk at the next meal. His fifty-year-old father drinks both fresh and not-so-fresh milk, and would eat the residue either at the first meal or on the second day. His mother is fine drinking the milk for a second meal, but never on a second day. According to Hui, 'It's hard to say (between mom and dad) who is more powerful (in my family). My dad is just thrifty, more open and more tolerant, more considerate.'

Ever since the one-child policy in China, feeding the country's little emperor became a tough task for the whole nation – privileging the best meal portions for children. But some parents still try to pass along food traditions to the next generation, though not always easily according to the description of one forty-year-old woman:

> It is always my husband who finishes leftovers at the next meal. But this supper is different because he is out, I will eat them instead. But today, I am trying to persuade my son to eat them. Once he saw the microwaved potatoes and spareribs, but he said he would not have them. I told him, 'We all know fresh ones are better but we "like" the remains. Now these are still fresh because it was cooked five hours ago. Your daddy always chooses to finish our leftovers first; actually he wants us to enjoy the fresh meals. That's silent love. Anyway, it's up to you ... Mom don't preach, ok? I will take it.' He did try some. Even after that I felt a little bit of regret, I should make sure my son only eats fresh meals since our life is better-off now.

Kids are not so passive in their involvement around thrifty practices. Mrs Sheng works and lives in the Pudong district, where high-tech international companies, banks and insurance agencies are concentrated. 'I'm glad that my eleven-year-old boy enjoyed the TV show *Extreme Cheapskates*. Though he does not like to eat dishes from the last meal, he may change in later years. At least, it helps him to understand that making money is not so easy, saving means a lot.' Today, children increasingly gain environmental information from viewing television shows, from kindergarten to primary school and college, formal or informal programs related to environment education are now part of school life.

Finish the leftovers, don't waste: from famine to sustainability

Most of the social practices related to food leftovers are based on daily routines, carried out unconsciously. People practice thriftiness or frugality, as part of a long tradition and old habit, which dies hard. Like the mother who encouraged her son to participate in eating leftovers, what they talk about relates to their family members, not to society or the world. In some particulars settings, the conversation of parents reaches a world far beyond the limit of the family. They are talking about people from other families, communities, provinces and countries, things from the past, even from decades ago.

Here are some cases where people are explicit about why they make use of leftovers and why they stick to these behavior patterns and persuade, suggest or encourage others to take similar actions. When asked, what is in your mind when thinking of or taking in food leftovers, the informants' responses are generally: 'Don't waste', 'Why waste?' and 'It is a (stubborn) habit'. In certain situations, some informants will state clearly the reasons why they make these statements, described in the three sections below and related to: learning from the past, caring for others, and environmental concerns.

Learning from the past: famine and food shortages

Fifteen-year-old Yokin moved to Shanghai, where his grandfather was born, with his father and mother nine years ago.

> If I wasted some food at the table, daddy often told me, 'You are lucky to have enough food to eat!' At one family reunion, my grandpa joined in. When they saw me throwing away half a piece of cake, he told me the story of *Adventures of Sanmao the Orphan*.[4] Sanmao had been hungry for two days. One day, he saw a gentleman feeding his dog with steamed meat dumplings. He was so hungry that he seized one from the floor. Before he was ready to eat, the dog became crazy and chased Sanmao. Frightened at the thought of being bitten, Sanmao had to throw back the precious food. This story shows how, in a time of food shortage, a boy's life was insignificant compared to that of a dog.

Adventures of Sanmao the Orphan is fiction, but the general picture depicted by the story itself is close enough to reality. Inner wars and the Japanese invasion brought about one disaster after another from the 1930s and 1940s in China.

More genuine table conversations relate the true history of modern China, as in the case of this thirty-five-year-old middle school teacher Mrs Dou, speaking of her father:

> My father likes drinking liquor very much. More often than not, when drinking with my father-in-law or other important relatives and friends on holidays and festivals, he talks about his experience of the famines during the 'Great Leap Forward', from 1958 to 1962. As he would say, 'It was hard to find roots of sweet potatoes in the fields, let alone green leaves. Tree skins were swiped for food. Leaves of poplar and elm trees were also taken away by hungry people. Some of our villagers starved to death. Several died and others were not strong enough to bury them properly. Some survived by stealing food from the commune and secretly keeping it for their family.' He turned to me and continued, 'It's really hard to remember this memory, but because of the food shortage, your grandpa and grandma left your baby aunt in the kitchen overnight, assuming the baby would die silently. Maybe because of the cold, the baby cried and was recuperated; with some warm water, she recovered. You could have not known your aunt, nor seen her now.'

This constitutes the collective memory of that generation, which involve discourses around what is simply termed 'the 1958 famine'. In the case of banquets or other important dining events, the memory of this famine is raised by men or women who will share their memories with others. In fact, hunger affected tens of millions of rural and urban Chinese during the tumultuous years of Japanese occupation and civil war from 1937 to 1949. There was a temporary improvement in the early 1950s, where the situation was stabilized through austere but adequate rationing, only to be swept away by the world's most destructive and largely man-made famine, which claimed at least 23 million and possibly as many as 30 million lives between 1959 and 1961 (Ashton et al. 1984; Peng 1987).

The other more recent and clearer collective experience is that of the hard times experienced during the planned economy. Rationed coupons were needed for all sorts of items, including the food necessary for survival, the cloth needed for piecing together clothing, or even the limited amount of industrial products available to urban and rural residents. Selling or buying agricultural outputs was forbidden, with free markets either disappearing or going underground. During this period of a nation-wide planned economy and rationing system, different provinces invented their own coupons, which could not be traded freely with other provinces. As all consumables were hard to find on the market, visiting friends or relatives in distant places meant bringing with you your own solid food or food coupons, so as to join in the local meals. In Shanghai, engaged fiancés could no longer afford a television set as betrothal gifts and waited for years to get a television-set coupon, through activating all possible social connections that might have access to these rare resources. This system began in 1955 and lasted for thirty years. For some of those who were born or grew up during this period, this life experience had a lasting affect on their attitudes and behaviors toward all kinds of material things and resources. A fifty-year-old college professor exclaimed:

> I never waste food. When I was in primary school, a few of my classmates were so starved that they fainted and could not walk back home. I went to a school six miles away. My father had to ask for help from a People's Commune cadre to trade our Jiangsu food stamps for national food stamps so that I could dine from the school's canteen. I hated that because our two families were not so close and my father had to do that for me. Sometimes some boys would not feel full after dinner, so they would go to nearby fields to steal carrots and sweet potatoes after class.

Caring for others

Beyond the experiences of the immediate family, some people relate to hardships experienced with food consumption and waste by neighbors, relatives or friends as a reference for dealing with food waste either left on the table or after disposal.

In my residential compound – one of the thousands of Shanghai's newly developed gated communities – Rui's grandfather does the cleaning for the community. In exchange, the property management has agreed to let him collect any discarded cardboard boxes or electronic equipment, for resale. He often complains about the hard work involved in handling the food waste, especially in the summer when it smells terrible. Mrs Liu, another neighbor, once told her daughter: 'You often play downstairs with Rui. Why do we dispose of so much food waste?', indicating that this was causing unnecessary trouble for Rui's grandfather. Her statement suggests that by becoming familiar with the work involved with managing waste by close friends or relatives, one can become more aware of the importance of leaving less waste in the rubbish bin.

To Mr Chang, his habit of being thrifty is not solely related to food shortage experiences in his childhood, but is also about his brother and sister and their experiences living in an unequal society. There are at least two situations that he found hard to forget and would recount with some difficulty. His younger brother moved to Shanghai as one of the many rural–urban migrants.[5] He opened a barber's shop, but the business was not sufficient for supporting his family. On his way back home in the late afternoon or early evening, he would occasionally go to a nearby food market to pick up Chinese cabbage leaves, which the vendors would not be selling the following day. He also saw older people picking similar perishable items from the market for their dinner, such as sprouted potatoes. Chang's older sister worked as a waitress in a restaurant in a Shanghai suburb. With her colleagues, she would often bring home chicken, pig or goose meat from the restaurant without the acknowledgment of her boss, so that her children could use these meats in meals they would prepare before or after school. Being sympathetic to them, Chang, now the owner of a small business and the richest among his cohort of friends, feels wasting food is out of the question.

The hardship experienced by people living in more undeveloped areas such as Yunnan and Guizhou are used as lessons when parents teach their children not to waste food. As forty-three-year-old technician Wu said to her son after her birthday party:

> You cannot spoil your food, do not discard your half bowl of dumplings in the bin. In Yunnan or Guizhou provinces there are some children who don't have enough food to eat, nor do they have enough clothes.

In the so-called information age, the lives of the poorest people in sub-Saharan Africa are also becoming a topic of conversation during mealtimes in China. One saleswoman in her late fifties told me about a conversation with her daughter:

> On the Lunar New Year Festival, my brother and sister came to visit us. This family reunion does not happen often. As usual we prepared more than enough food, not only to impress them but also as a gesture of hospitality. Five or six dishes of food are still there. We will not waste them. When my daughter asked why we would not throw the leftovers away, I told her, 'Remember the TV programs of African life? So many children suffer so much from malnutrition that they seem to have only bones left on their bodies. Thousands die before they can grow up to be your age. You have no reason to waste food.'

Environmental concerns

There are a growing number of people who handle food with serious environmental concerns. Expressing one's opinions publicly is not popular in Shanghai, as a sense of community citizenship has declined in the city, with rising concern for privacy. Most people would be very reluctant to express themselves frankly to strangers, and even hesitate to remind their friends, colleagues or relatives to be more cautious with regards to environmental issues. Such discussions are common, however, among family members and people with close social ties. During the interviews on food waste, I heard informants express environmental concerns several times. As one person said: 'Water and land pollution is so serious. It's a crime to waste. Money in your pocket is yours, but resources belong to the world.' Mr Ma gained his environmental consciousness from relatives living in Japan:

> From one relative living in Japan, I know that the Japanese do a good job in terms of waste management. When talking with Japanese friends, I realized that municipal solid waste has been a problem for every city in the

world. And wasted food is one of the most important aspects. It is also one of toughest to process, in regards to disposal, collection, treatment etc. Regarding the environment, nobody likes the rubbish on the grass or inside parks. So why would I dispose of heavier food waste?

Besides finishing food leftovers, these inconspicuous environmentalists find other ways to embrace a green lifestyle. While some people discard certain parts of a vegetable, others use them in family dishes:

> There is a routine dish in my family. Every time we buy celery, besides the celery stem, we make use of the leaves to make another dish: fried eggs with celery leaves, or soup of celery leaves, eggs and mushroom. The vegetable itself is green, so these dishes really give us the impression that we are living green. By consuming the leaves, we don't throw them away. The waste management staff will collect less waste and less energy will be used to process this waste.

Sixty-year-old Zhao goes even further in his use of peelings: if the watermelon peel is thick enough, he makes use of it in a dish, especially early summer when other vegetables are not so rich and varied. He even shared his experience and skill with his sisters-in-law: 'You have to stir-fry watermelon peel on a high temperature fire to make it tasty,' he explained.

Hui's mother uses the residue from beans after making soya-bean milk for her plants. Rather than consuming it – as her husband would do, saying, 'It's mine, it's no problem and I would like to have it' – she prefers placing the residues in her flower pots. 'Here as the fertilizer, I still reuse and recycle it. And you don't have to consume something controversially nutritious.' She recalls her parents making use of such waste as organic fertilizer in the farming fields of their home town, Anhui, and this remains a vivid memory that she has carried forward into her life in Shanghai.

In recent history, many Chinese people were too poor to afford meat. During the past three decades, however, staple grains have become less desirable and the intake of animal foods has increased, pushing up demand for high-protein feeds. Meat consumption has been rising faster than anticipated: it close to tripled between 1978 and 1994. Mainly out of health concerns, people over the age of fifty are becoming more interested in consuming vegetarian foods rather than meat. Mr Qiu and his wife are over sixty years old and state their vegetarian preference as follows:

> My wife and I are becoming more vegetarian for several reasons: almost all of the pigs, chickens and cows are raised on commercially manufactured feeds, which can fatten the animals up quickly enough for market value. Eating too much of these meats is harmful to health. People of our age suffer from obesity, high blood pressure, hypertension, etc.

Like Hui's mother, Qiu added, 'If there is some vegetarian food waste, we will spread it in our garden to make fertilizer for the plants and flowers.'

If some people do not clearly express environmental concerns, which are unconscious, others are conscious and intentional practitioners of pro-environmental actions. Mrs Tian is the leader of a non-governmental organization (NGO) focused on improving the living conditions of school children in underdeveloped areas of China. When gathered around a table for a meal, she encouraged us with the following statement:

> There are many reasons to go vegetarian if you care about the environment. I have found out that so many environmentalists are vegetarian. It's becoming common sense for those of us with a high educational background that raising animals for food requires massive amounts of land, food, energy and water. The byproducts of animal agriculture pollute our air and waterways. I'm trying to have less and less meat every time I think of this. Regarding food leftovers, vegetarian food is easier to prepare and finish (than meat).

The case of wedding banquets

When asked which situations will cause significant food waste for most people, informants claimed that wedding banquets are the main culprits, among all the important events and celebratory parties. There are also lots of food leftovers and food wasted during holidays, such as the Spring Festival, New Year's Day, National Day, Dragon Boat Festival or Mid-autumn Festival. Socially important events and parties often take place during such holidays. In one focus group discussion I organized with acquaintances, participants began discussing how to deal with dining out and food leftovers. Mr Li works in a law firm as a solicitor and his personal experience as a wedding banquet participant aroused enthusiastic discussion:

> For one wedding ceremony, thirty-two banquet tables were prepared in that grand restaurant. Soon after the majority of the invited guests were gone, I noticed that the senior relatives from the bridegroom's side asked for a dozen plastic storage bags and collected the remaining braised pork, fish, shrimp, chicken, pig knuckles etc. I suddenly realized what they were doing. Many of these food items were already touched by guests' chopsticks. They didn't even seem to mind that a few of us were still there, although in fact we were quite set with food.

Mr Yang, whose family have been residents of Shanghai for more than five decades, responded in the Shanghai dialect, addressing various habits of China's different regions:

> It's no surprise. Ten years ago I visited North East China. It is fashionable there to entertain and impress their guests with a huge table of dishes. The

more leftovers, the more hospitable and enthusiastic is the host. As a result, half of the food was left on the table. Too much waste! We Shanghainese don't show off in this silly way. At such banquets, we do organize close relatives to pack (the leftovers) and invite their own relatives and friends to share in the reheated food the next day. This custom here is named 'eating tablecloth'. All Shanghainese do this and people living in other places may also have this habit, but in Shanghai's restaurants and hotels, it is common for leftover packages and bags to be readily available.

All participants in this focus group are in favor of packing the leftovers, though they emphasize different aspects. Miss Yang, a government office clerk, linked this practice with food culture:

This is nothing wrong. This is culture. It has nothing to do with being rich or poor. My boyfriend's uncle owns a prosperous company and their family always packs their food leftovers (after dining out).

Mrs Ma's response demonstrates one feature of what it means to be Shanghainese today: a combination of a reasonable, mild or strong sense of personal rights in the commercialized metropolis, along with sensitivity to resources and environmental problems:

That's very good. You pay for the meals, so the leftovers are also yours. It's your right to let your relatives pack and share. Or the waiters will remove all to the dustbins. That's really a huge waste and no good for *Huanbao* (environmental protection).

While being thrifty and saving is praised, waste is despised, though the value judgment is not always apparent, as in the case of Mrs Sun: 'It does not matter how expensive it is to dine out, but from the bottom of my heart I always despise those who waste too much.'

Mr Yu's comment placed an emphasis on family background, social class identification and taste culture, especially in relation to local residents:

I heard of the saying that a businessman should always leave a mouthful of food and should not turn over his fish[6] to show their upper-class exquisite family training. But to be frank, that is false pride. Real Shanghainese never have such rules. Shanghainese have always been frugal. They are very good at making use of everything. Thus they are often mocked as 'stingy', 'overly calculating' or 'shrewd'.

Mr Yang also connects the situation with the popular image, especially from new social media, of how American and European celebrities finish all the leftovers on their plates:

It is an indication of moral quality for *Yang Daren* (important foreigners) to clean their plates. Why shouldn't we? China has the tradition of being frugal and hardworking. China has now switched to an international track and that habit should be popularized.

Conclusions and discussions: towards strong sustainable consumption in China

With the world's largest population, a long agricultural history, limited natural resources, increasing pollution and the collective memory of serious famines, pathways towards more sustainable forms of consumption is a key area of research and action in China. Regarding food waste, the handling of food leftovers in China often involves strong feelings of love and intimacy among very close social ties, as revealed by the Shanghai example put forward in this chapter. Based on my research, most people do practice thrifty and frugal food consumption, which leads to both food sharing and reduced food waste, and few people would choose to waste food, regardless of whether the food was private or collectively funded. That being said, socially and politically significant events still go hand-in-hand with extravagant banquets, oftentimes paid for by public funds, and associated with food waste. Eating out is therefore a key area of study for food waste in urban contexts, as emphasized in other contributions to this book. Regarding finishing food leftovers or packing leftovers from restaurants, most people take these actions for the welfare of their own family, but a growing number clearly state reasons other than family care. Some people have personal experiences of food scarcity and famines, some express concern for others who handle food waste, while a growing number of people seem to be motivated by environmental concerns, often expressed in terms of 'environmental protection' rather than 'sustainable consumption'. For people with personal experiences of food scarcity, these habits are stubborn and hard to change, with concerted efforts underway to educate their children to do the same. It is not clear whether future generations, who may not have memories of food famine and shortages, will retain a concern for not wasting food.

All means are valid towards increased sustainable food consumption in China, yet the notion of thriftiness may not be an adequate long-term goal towards greater sustainability. As Evans has argued (2011), there is a distinction to be made between thrift and frugality: the former entails saving money on certain items, but with no limits to overall consumption; the latter is a focus on moderate consumption patterns overall, or careful consumption and the avoidance of waste. Based on empirical research among households in London, Evans suggests that frugality challenges normative expectations around consumer culture and could be a better approach to environmental sustainability, rather than thriftiness. There is a generational gap when it comes to motivations for reducing food wastage in Shanghai: for older generations, food is not wasted due to values that embrace restraint over indulgence, in addition to growing health concerns. Rather than 'thriftiness', the more appropriate term could be 'frugality', in terms of caring about scarce resources rather than the cost of consumption. Among younger generations, and particularly young professionals, the resistance to food waste is more on an environmental register. Here, the discourse and actions of young people mirror the tensions apparent in the political agenda for China, which faces a development strategy dilemma: on the one hand, political discourse calls for economic growth by means of expanding consumption; on the other hand, the government promotes a circular economy and, for long-term sustainability, greener lifestyles, a cleaner environment and ecological citizenship would all be necessary. The consumer revolution that has taken place all over China since the economic reforms has granted people the freedom to satisfy their depressed desire to live better lives, yet Chinese people must not only resist the lure of spreading consumerism, but also the government's call for increasing consumption to promote economic growth, even if this entails more pollution by products 'made in China'.

Facing these contradictory messages and pursuits, the differentiation between what motivates reduced food waste among everyday people is significant. For a great number of people living in China today, the lure of consumerism is strong and environmental concerns are simply at the front of the mind. Even if most people avoid food waste, overall trends in consumption promote what could be seen as wastefulness, such as the desire for bigger houses and cars and the frequent purchase of an increasing number of superfluous things – all signals of unsustainability. The cheaper the better for many living in China today, which tends to enforce a notion of thriftiness rather than frugality. For others who are sensitive to environmental issues, one conceivable way forward would be to shop their way towards green consumerism, a trend that has been noted in the context of Europe and North America, which is also not sustainable over the long term. Older generations in Shanghai retain a sense of frugality and a moral opposition to over-consumption and waste, which could entail a stronger trend towards sustainable consumption. Yet the memory of food shortages driving this behavior could dissipate over the years and among younger generations.

The Chinese have a long way to go, from caring for themselves and the welfare of their own family, to a sense of responsibility towards global sustainability. This touches on issues relevant to sustainable consumption, such as ethical consumption, the role of citizen-consumers, and the significance of lifestyle change – all growing areas of research in academia in the northwestern hemisphere (Spaargaren 2003). More research is needed in the context of Asia and the Pacific on consumption patterns and practices, with this volume as a first step in that direction. In the case of Shanghai, conspicuous food consumption remains an important trend, yet the handling of food waste is tied to notions of frugality for some. Based on this research, I would argue that, over time, health aspects and notions of altruism, such as caring for family and for others, could be further reinforced. Through these two themes, tied to notions of frugality, people in Shanghai and possibly in other urban contexts in China could transition from thrifty consumers to consumer-citizens, towards more sustainable lifestyles.

Acknowledgements

The research presented in this chapter was made possible with support from the National Social Science Fund of China (15ASH009). My thanks go to Marlyne Sahakian, Suren Erkman and Czarina Saloma for their reviews and editing.

Notes

1 http://www.worldwatch.org/food-waste-and-recycling-china-growing-trend-1
2 All of the interviews were conducted in Chinese and were translated into English by the author.
3 Tan Qiuming. 2015. 'Refusal of eating overnight food leading to divorce.' (In Chinese) http://newsapp.gzdaily.com/jsp/share.jsp?code=NjQ1Njc= (accessed 6 November 2015).
4 Sanmao, which means 'three locks of hair' in Chinese, is traditionally a popular nickname for children, especially around the Yangtze Delta. Vividly depicted by Zheng Leping's popular children's book *Adventures of Sanmao the Orphan*, this poor boy's bitter stories have moved millions of readers around the world.
5 China's rural–urban migrant workers may constitute the world's largest migration. These labor-age peasants migrate to urban areas to work in factories, on construction sites, in mines, in agriculture, in producer services (security guards, cleaners) and as small-level self-employed (in shops, in markets, as scavengers). Their most important aim is earning money and saving money, partly to take care of their parents or children left at home in their own village. Their working and living conditions in the city are precarious. Food and housing rent are dominant parts of their expenditure; most of them try hard to cut down on both.
6 The image of a fish turned over reminds people of a shipwreck, a popular metaphor in south China for bad luck in business, or even bankruptcy.

References

An, Y., G. Li, W. Wu , J. Huang, W. He and H. Zhu. 2014. 'Generation, collection and transportation, disposal and recycling of kitchen waste: A case study in Shanghai.' *Waste Management & Research* 32(3): 245–248.
Ashton, B, K. Hill, A. Piazza and R. Zeitz.1984. 'Famine in China, 1958–61.' *Population and Development Review* 10: 613–645.
Bourdieu, P. 1977. *Outline of a Theory of Practice*. Cambridge: Cambridge University Press.
Buck, J. 1937. *Land Utilization in China*. Nanjing: University of Nanking Press.
Cappellini, B. 2009. 'The sacrifice of re-use: The travels of leftovers and family relations.' *Journal of Consumer Behaviour* 8(6): 365–375.
Chang, K. 1977. *Food in Chinese Culture: Anthropological and Historical Perspectives*. New Haven, CT: Yale University Press.
Evans, D. 2011. 'Thrifty, green or frugal: Reflections on sustainable consumption in a changing economic climate.' *Geoforum* 42: 550–557.
Farrer, J. 2011. 'Globalizing Asian cuisines: From eating for strength to culinary cosmopolitanism.' *Education About Asia* 16(3): 33–37.
Giddens, A. 1984. *The Constitution of Society*. Cambridge: Polity Press.
Li, L. 1982. "Introduction: Food, famine, and the Chinese state.' *The Journal of Asian Studies* 41(4): 687–707.

Mallory, W. 1926. *China: Land of Famine*. New York: American Geographical Society.
Peng, X. 1987. 'Demographic sequences of the Great Leap Forward in China's provinces.' *Population and Development Review* 13: 639–670.
Pye, L. 1981. 'Foreword.' In *Shanghai, Revolution and Development in an Asian Metropolis*, edited by C. Howe, pp. xi–xvi. Cambridge: Cambridge University Press.
Reckwitz, A. 2002. 'Toward a theory of social practices: A development in culturalist theorizing.' *European Journal of Social Theory* 5(2): 243–263.
Shove, E. 2003. *Comfort, Cleanliness and Convenience: The Social Organization of Normality*. Oxford: Berg.
Smil, V. 1995. 'Who will feed China?' *The China Quarterly* 143 (Sep): 801–813.
Spaargaren, G. 2003. 'Sustainable consumption: A theoretical and environmental policy perspective.' *Society and Natural Resources: An International Journal*, 16(8): 687–701.
Wang, X., Q. Wang, Y. Liu, and H. Ma. 2010. 'On-site production of crude glucoamylase for kitchen waste hydrolysis.' *Waste Management & Research* 28(6): 539–544.
Ward, R. and J. Li. 1993. 'Solid-waste disposal in Shanghai.' *Geographical Review* 83(1): 29–42.
Warde, A. 2005. 'Consumption and theories of practice.' *Journal of Consumer Culture* 5(2): 131–153.
Welch, D. and A. Warde. 2015. 'Theories of practice and sustainable consumption.' In *Handbook of Research on Sustainable Consumption*, edited by L. Reisch and J. Thøgersen, pp. 84–100. Cheltenham: Edward Elgar Publishing.

10 Food waste in the food service sector

A case study from Kuala Lumpur, Malaysia

Effie Papargyropoulou

Every year one-third of all food produced for human consumption is lost or wasted (FAO 2013). This wastage represents losses of valuable nutrients, energy and natural resources. At a global scale food wastage is also responsible for greenhouse gases (GHG) emissions comparable to those of the top ten GHG emitting countries in the world (FAO 2014), as well as contributing to food insecurity and posing ethical questions. How is it that the global food system leaves one in nine people malnourished or hungry and on the other hand wastes one-third of the food supplies? Food wastage represents an annual loss in economic value equal to USD 1 trillion across the global food supply chain (FAO 2013). The magnitude and complexity of the food waste problem has brought it to the forefront of the political agenda in most developed countries. This problem is expected to continue to grow, especially in developing countries, considering the likely changes in the food systems in these countries due to industrialisation, urbanisation and changes in diets and lifestyles (Garnett and Wilkes 2014).

Academic research on food waste is emerging. However, it mainly focuses either on the macro level, examining waste management policy, waste treatment technology and food wastage in agriculture and retail, or on the micro level of the household. The meso level remains understudied and this is exactly where this study is situated. This chapter presents a case study of food waste generation at a restaurant in Kuala Lumpur, Malaysia. The aim of the study is to investigate the mechanisms of food waste generation and explore opportunities for food surplus and food waste prevention. The case study is located in a developing country experiencing rapid economic growth, changes in food consumption patterns and associated social practices. Such a dynamic context makes for a unique contribution to the academic literature on food waste. A secondary aim of the study is to illustrate the strength of interdisciplinary research when studying the complex phenomenon of food waste, by demonstrating how only through an interdisciplinary approach can we fully connect the social and material contexts within which food waste occurs.

Malaysian context and food waste background

The case study presented in this chapter is located in Kuala Lumpur, Malaysia, a country that aims to reach high-income status by 2020. In this context, urbanisation is seen as the driving force to achieve this aspiration and Kuala Lumpur is central to Malaysia's urbanisation strategy, tasked with increasing labour productivity and the average household income by a factor of ten by 2020 (Pemandu 2012). These changes have a substantial impact in shifting food consumption practices and, as a consequence, patterns related to food waste. Eating outside the household is becoming increasingly popular as manifested by a thriving food service sector (Malaysian–German Chamber of Commerce 2012) and new trends such as buffet-style restaurants are gaining popularity and are becoming the norm. The emerging patterns observed relate to the coupling of food waste generation with the economic growth experienced (Johari et al. 2012) and as a result the doubling of the per capita food waste generation in the last thirty years (Agamuthu et al. 2009).

The relationship of Malaysians with their waste, and in particular their food waste, can be described by a sense of detachment. The emergence of municipal solid waste management made food waste invisible to the everyday waste producer around the world (Evans et al. 2013), however food waste in Malaysia has yet to regain visibility in political and cultural life as it has in the developed world. The invisibility of food waste is even more exaggerated when food consumption takes place outside the household. As a result, the food waste producer often underestimates how much food waste is actually produced (Waste and Resources Action Programme 2007a, 2007b). In the context of a developing country such as Malaysia, often food waste is seen as a sign of affluence and development, and the individual feels it is the role of the state to address the issue – through waste management – and any individual responsibility is relinquished.

Although food waste has substantial negative economic, environmental and social impacts throughout the food supply chain (Gustavsson et al. 2011; Papargyropoulou et al. 2014; Padfield et al. 2012), it is a peripheral or insignificant issue in policy debates in Malaysia, as are broader issues related to sustainable consumption and production. Food waste is only perceived as a potential resource for energy generation through biological and/or thermal waste management technologies. Hence, awareness of sustainable food consumption, food waste prevention and minimisation, and food surplus redistribution to groups in need is limited amongst Malaysians.

Food waste definitions

In this study, food waste refers to food which was originally produced for human consumption but was not consumed by humans; instead it was directed into a non-food use (for humans), feed for animals or waste disposal (for example, feedstock to an anaerobic digestion plant or incinerator, disposal at a landfill) (FAO 2014). Avoidable food waste refers to food that could have been eaten at some point prior to being thrown away, even though much of it would have been inedible at the point of disposal (Quested et al. 2011). Unavoidable food waste refers to the fraction of food that is not usually eaten, including items such as banana skins, apple cores, egg shells and bones.

Methodology

A mixed methods interdisciplinary strategy was developed for the collection and analysis of both qualitative and quantitative data (based on Papargyropoulou et al. 2016 and described in detail below). By combining qualitative and quantitative methods and approaching the research from more than one angle, the mixed method strategy embodied the notion of triangulation (Jupp 2006). The proposed methodological approach was based on an inductive and iterative process in which theory was built and modified based on the data collected. The constant comparative method was applied by continually comparing sections of the data to allow categories to emerge and for relationships between these categories to become apparent. The emerging categories were then modified into more abstract concepts. Theory was built by organising these concepts into logical frames. As new data continue to emerge, new concepts were added until a point of 'saturation' was reached whereby new data did not contribute anything new to the theory (Glaser and Strauss 1967).

The overarching research method was that of a case study, using in-depth and semi-structured interviews, focus groups, observations and quantitative data collection techniques. As a research strategy, a case study involves the empirical investigation of a particular contemporary phenomenon within its real-life context, using multiple sources of evidence (Saunders et al. 2009). The unit of analysis for the case study was a restaurant within a five-star hotel in Kuala Lumpur, Malaysia, offering the opportunity to study how several variables (including type of food service such as buffet and *à la carte*; type of customer; time of meal such as breakfast, lunch or dinner; type of cuisine such as Asian, Malaysian or Western) influence food waste generation. Within the unit of analysis, the phenomenon of food waste generation was studied from the time of purchase of raw food supplies, throughout food storage, preparation and cooking, customer consumption, and finally, discarding of food waste. It did not include waste collection and final disposal at the landfill or other waste treatment facilities. Data collection was carried out over three months in May, June and July 2014.

Ethnographic and qualitative methods: interviews, observation, and focus groups

Two types of interview were carried out in this study: in-depth and structured, and informal and non-structured. The interviewees included both employees from the case study restaurant and other stakeholders such as representatives of the National Solid Waste Management Department. In-depth interviews of key actors were carried out in order to understand the broader context in which food waste generation occurs in the food service sector. Following the initial round of interviews, participant observation combined with informal discussions with the employees was carried out while collecting quantitative data. The observations were recorded through field notes in the form of a diary. An important advantage of participant observation was that it provided a form of triangulation for the other research methods, such as the interviews and the quantitative data collection methods. A focus group discussion was also carried out following some preliminary data analysis. The main patterns emerging from the data were discussed in the focus group comprising members of the management, procurement, sales, finance, food preparation and operations teams. This allowed further analysis and opportunity to seek clarification on behaviour recorded during the participant observation. It offered further insights as to where, how and why food waste is produced and what can be done to prevent it.

Quantitative data and tools from industrial ecology

The quantitative data collection methods aimed to identify processes and activities within the restaurant that give rise to food waste. The goal was to measure the amount of food waste generated from these processes in order to provide evidence vital to prioritising the most promising measures for waste prevention. The quantitative data collection methods comprised of a food waste audit, the collection of photographic records, financial records and an inventory of food purchases. The food waste audit was carried out over one week, during which food waste was weighed during breakfast, lunch and dinner for seven consecutive days, from 9am to 12pm. A total of 1205 meals were served during the waste audit. Six kitchens/food preparation areas and their respective washing areas were monitored. During the food waste audit, the amount and type of food waste were identified. The amount of food waste generated was measured and recorded continuously throughout the day.

Three types of food waste were recorded. 'Preparation waste' refers to food waste produced during the food preparation stage, due to overproduction, peeling, cutting, expiration, spoilage and overcooking, among others. 'Customer plate leftover waste' refers to food discarded by customers after the food had been sold or served to them. 'Buffet leftover waste' refers to excess food that had been prepared but had not been taken onto the customer's plate or consumed, and thus left on the buffet or in a food storage area and discarded later on. The ingredients of the food waste were also recorded and categorised into nine food commodity groups including cereal, dairy, eggs, fish and seafood, fruits, meat, oils and fats, sauces, and vegetables. This allowed for detailed material flow diagrams. This information was then combined with the food-purchasing inventory, to produce economic flows and eco-efficiency ratios for each food commodity group. Sankey flow diagrams were used to visualise the magnitude of economic and material flows taking place within the case study. The thickness of each link represented the amount of flow from a source to a target node, in this occasion from food provisioning to food consumption. In addition to the amount of food waste generated and the process that gave rise to it, in-situ estimates of the avoidable and unavoidable fractions of food waste were made based on visual observations.

Strengths and challenges of the methodological approach

The merit of this methodological approach lies in the fact that it links the social practices related to food consumption with the quantitative patterns and biophysical flows of the food waste phenomenon. It provides valuable insights and empirical data into food consumption and food waste generation in the food service sector in a developing country. This case study does not aim to provide definitive data on food waste generation applicable across the whole food service sector. Repeating this study in other settings would further improve and refine the proposed methodology and provide data on food waste generation more widely applicable. The main challenge of the methodological approach presented is that it is labour and time intensive. It is this intensiveness, however, that provides the depth and richness in data analysis required to understand food waste generation.

Case study results and findings

Hotel restaurants are becoming increasingly popular among diners; thus the choice of a restaurant operating in a five-star hotel for this study. The hotel belongs to a prominent high-end international hotel chain catering for a mixture of foreign and local tourists, and people travelling on business. The study focused on the main restaurant of the hotel that served breakfast, lunch and dinner. Breakfast was in the form of a buffet and catered primarily for the hotel guests, although walk-in customers were also accepted. Lunch service was in the form of a buffet between Monday and Saturday, and *à la carte* menu every Sunday. Dinner service was in the form of *à la carte* with the exception of Saturdays when special buffet events were organised. The restaurant offered a mix of Malaysian and Western cuisine. The buffet meal cost RM 126 (USD 35) and an average meal from the *à la carte* menu cost approximately RM 80 (USD 22). The restaurant's operating hours were 6.30am to 11.00pm, Monday to Sunday.

During breakfast, customers were predominantly hotel guests, a mixture of local and foreign tourists, and business people, with significant disposable income, spending a minimum RM 500 (USD 140) per room per night. Customers during lunchtime were largely local residents and a few foreign hotel guests. The local lunchtime customers were predominantly high-income government workers from the neighbouring government administrative area, as well as business people on lunchtime meetings. Customers having *à la carte* dinner were mostly hotel guests, although the Saturday night buffet attracted customers other than hotel guests.

The amount of food waste generated by the restaurant is presented in Figure 10.1 below. Food waste was broken down into preparation waste, buffet leftovers and customer plate leftover waste. On average 173 kg of food waste per day or 1212 kg per week was generated by the restaurant's operations. Extrapolating from this average daily rate of food waste generation suggests that in one year this restaurant would generate in the region of 63 tonnes of food waste.

A noticeable daily variation in amount of food waste can be observed. However, this did not always correlate to the number of customers served per day. For example, Tuesday appeared to have the second highest amount of food waste, although more customers were served on Saturday, Monday, and Thursday (see Figure 10.1). This apparent anomaly can be explained by the fact that part of the food preparation (and subsequently generation of preparation food waste) occurs on the day before, not on the actual day of a given event (e.g. on Tuesday some preparation was made for Wednesday's buffet, which had the highest number of customers).

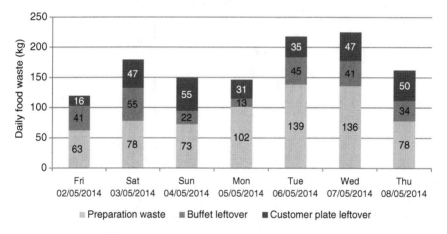

Figure 10.1 Daily food waste generation.

Table 10.1 Daily number of customers and food waste per customer rates

	Fri 02/05/14	Sat 03/05/14	Sun 04/05/14	Mon 05/05/14	Tue 06/05/14	Wed 07/05/14	Thu 08/05/14
Customers per day	101	168	89	161	148	295	243
Food waste per customer (kg/person)	1.2	1.1	1.7	1.0	1.5	0.8	0.7

Another example is Sunday, when the highest daily food waste generation per customer was recorded (1.7 kg per customer, see Figure 10.1). Further analysis revealed that on Sunday preparation waste per customer was the second highest recorded that week (0.8 kg per customer, see Figure 10.3), in particular during lunch and dinner times when *à la carte* service was offered (as opposed to buffet service). This suggests that *à la carte* service produced more preparation waste per customer compared to buffet service. In addition, customer plate waste during lunchtime was the highest recorded that week (1.37 kg per customer). Qualitative methods, such as observation of food consumption practices and informal discussions with staff and customers, were used to explain the quantitative patterns observed. In more detail, on Sunday only one family of seven had *à la carte* lunch. Discussions with staff revealed that the family were tourists from a Middle Eastern country on vacation in Malaysia. According to the waiter that served the family, the leader of the family ordered food above what was required for seven people:

> Waiter: 'He ordered too much, you know for seven people only, three pizzas, seven portions of *nasi* (rice), three whole chickens, starters, salads, bread, too much…'
>
> Researcher: 'Did you tell him it was too much? Did you advise him on the portion sizes?'
>
> Waiter: 'Yes of course, but you know with customers you can't insist too much, they are the customers. Also in some cultures the man has to provide for his family, his wives and children, and show he can buy more than they need. I see that thing again and again, Europeans will order enough but not too much. The Japanese are the same you know, they only order what they need and they don't leave anything on the table. This guy ordered seven desserts afterwards and half of the food on the table was not even touched. It's not right you know, but we can't do anything about that.'

This is one example of many encountered in this study, where perceived cultural differences were given as the reasons behind certain consumption practices, wasteful or otherwise. In addition, it shows the anxiety food waste causes (for anxiety associated with food wasting in the household see Evans 2011); in this case not to the waste producer but to the waiter feeling uncomfortable with the wasteful practices of the customer.

Observations of food consumption in the buffet setting also highlighted the link between food waste generation, in particular customer plate food waste, and the customers' perceptions of 'value for money'. Discussions with customers and staff revealed that the notion of 'value for money' closely relates to quantity not necessarily quality of food. A typical response when asked whether a customer was satisfied with the buffet is presented in the dialogue below:

> Researcher: 'Are you satisfied with the buffet? Did you feel it was worth the money you paid?'
>
> Customer: 'Yes, the buffet is good value for money. It has so many items, a lot of choice, and the buffet was full even towards the end. I tried them all.'

Quality and taste of food, standard of service, and other factors were rarely mentioned in relation to value. A large variety of food items to choose from was an important factor in determining customer satisfaction. Maintaining the same level of choice (that is, the buffet not running out of food items) even towards the end of the mealtime was another important factor quoted by customers. The restaurant management and staff were aware of this:

> Head chef: 'We need to offer the same choice to the last customer as the first one. They pay the same money, so we need to give them the same service. That's why our buffet is always full even five minutes before it closes. If that means we have to waste a tray full of food that hasn't been touched, then that's what we need to do.'

These examples demonstrate how food waste generation was affected by the type of service provided, for example *à la carte* or buffet, and expectations and food consumption practices of the customer, as influenced by perceptions and social norms.

An important factor to consider when quantifying food waste generation was how much of the food waste was actually avoidable (for definitions refer to the 'Food waste definitions' section above). Figure 10.2 illustrates the percentages of avoidable and unavoidable food waste within this case study. In total 56 per cent of all food waste generated was avoidable, which shows the significant scope for food waste prevention. At the preparation stage, the majority of food waste was unavoidable as it was comprised of mainly inedible parts of foods, such as bones, seafood shells, inedible fruit skins and cores, among others. Buffet leftovers were mainly edible, thus the avoidable fraction was 94 per cent. Food waste from the customers' plates was a mix of inedible parts such as bones and seafood shells, and edible surplus food. The overall unavoidable fraction measured in this case study (44 per cent of total food waste) is significantly higher than the one Betz et al. (2015) report (maximum 21 per cent unavoidable fraction). This is due to the nature of the restaurant in this case study: the restaurant prepared meals using fresh ingredients and few processed items. This resulted in having all the preparation waste associated with a certain meal, produced within the restaurant and not in previous stages of the food supply chain, such as in food-processing industries. The high preparation waste consisted of inedible parts such as bones and exotic fruit skins, for example. The second reason is that in this study possibly avoidable food waste was reported within the unavoidable fraction. These type of variations, due to the subjective nature of definitions of avoidable and unavoidable fractions, and due to the extent by which the restaurant uses pre-prepared food, have also been highlighted by Betz et al. (2015).

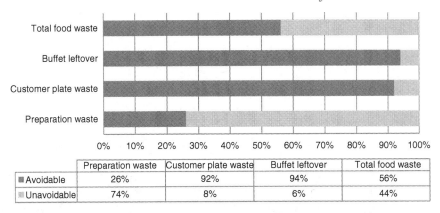

Figure 10.2 Avoidable and unavoidable food waste as a percentage of total food waste.

Figure 10.3 shows the average food waste generated per customer served. These figures can act as a benchmark for food waste generation, regardless as to whether many or only a few customers were served at a particular time. These figures suggest that the lunchtime *à la carte* meal had the highest food waste generation rate. However, this figure was based only on one meal time (Sunday 4/5/2014), which was a particularly wasteful occasion as discussed above. The breakfast buffet had the second highest food waste generation rate at 1.2 kg per customer served, followed by the lunch time buffet with 1.1kg per customer and dinnertime buffet and *à la carte* service, with 1 kg per customer. If the outlier of the lunch time *à la carte* meal is excluded, the figures suggest that buffet-style service was overall more wasteful than *à la carte* service. Buffet service had lower preparation waste per customer rates as explained by economies of scale; however it produced substantial amounts of buffet leftovers, making it a more wasteful type of service.

Based on qualitative observations, the food items most commonly wasted were rice and noodle dishes, cakes and desserts, and fruits. Rice and noodles contributed considerably in both buffet and customer plate leftover waste. Rice also contributed to preparation waste, as there were instances of overproduction and rice stuck to the bottom of cooking pans. Cakes, desserts and fruits contributed significantly to food waste generation in the form of buffet leftovers. Fruit waste also appeared in the form of preparation waste for a number of reasons. Skins and cores from fruits such as watermelons, melons, mangos and pineapples are substantial in weight, therefore they contributed to the high unavoidable preparation waste rate. Avoidable preparation waste was generated when fruits were used as part of plate decoration and elaborate designs, which required large edible parts of the fruit to be wasted in order to achieve the desired shape. Another example of avoidable preparation waste was bruised, spoilt and overripe fruits not used on time. These observations corresponded with the results of the material flow and eco-efficiency analyses presented below.

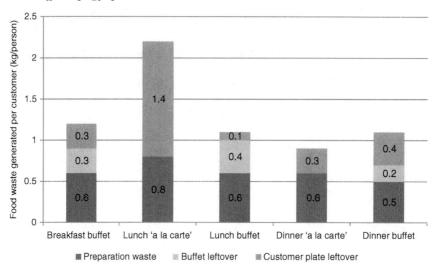

Figure 10.3 Average food waste generation per customer served for different mealtimes.

Sankey flow diagrams were used to visualise the magnitude of economic and material flows taking place within the case study. The thickness of each link represented the amount of flow from a source to a target node, on this occasion from food provisioning to food consumption. Three Sankey diagrams are presented. Figure 10.4 shows the material flows and breakdown between food consumed, preparation, buffet leftovers and customer waste. On the left-hand side of the figure 'incoming food' represents the mass of all the food purchased by the restaurant. According to the analysis of incoming food and the outgoing food waste, it was calculated that approximately 30 per cent of purchased food was lost in the form of food waste. In more detail, approximately 17 per cent of food was lost during preparation, 7 per cent as customer plate waste and 6 per cent as buffet leftover waste. The total food waste rate is higher than the average reported by Beretta et al. (2013) at approximately 20 per cent, however lower than the maximum food loss they encountered during their study, of 45 per cent at a gourmet restaurant. In Figure 10.4 the liquid fraction was included within the incoming food, food consumed and food waste, rather than being shown separately.

Figure 10.5 presents the material flows in terms of nine commodity groups. Meat and dairy represent 10 per cent and 8 per cent of incoming food; however, only 1 per cent and 0.2 per cent of these food commodities respectively left the restaurant in the form of waste. However, vegetables, cereal and fruit represented the three most wasted food commodities. This finding corresponds with reports by other studies (Al-Domi et al. 2011; Betz et al. 2015).

Figure 10.6 shows the economic flows taking place within the restaurant, broken down in the nine food commodity groups. This graph provides a different perspective as compared with the previous graphs. It shows that

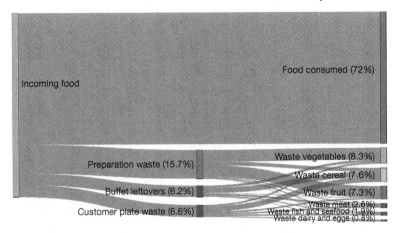

Figure 10.4 Material flows.
Software by Bostok, 2014.

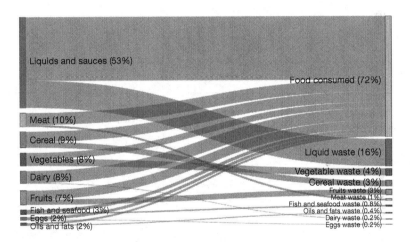

Figure 10.5 Material flows in terms of food commodities.
Software by Bostok, 2014.

despite the fact that the liquid fraction was the most significant waste component in terms of weight (55 per cent of total waste), it was not significant in economic terms. In contrast, cereals, vegetables, fruits, fish and seafood were the biggest economic losses of the system.

In order to calculate the eco-efficiency of the different food commodities, the cost value was matched with the environmental value, in this case waste generation. Cost value was expressed in Ringgit Malaysia (RM)/kg of food, and the environmental value as a percentage of food wasted. Figure 10.7 compares the food commodities in terms of their eco-efficiency. Cereals, fish and seafood appear at the top right quarter of the graph, representing food

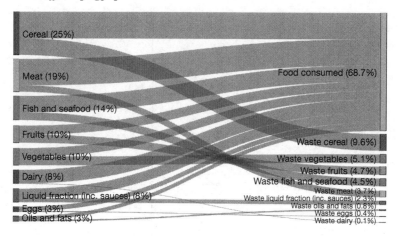

Figure 10.6 Economic flows.
Software by Bostok, 2014.

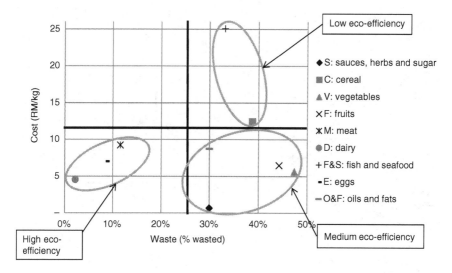

Figure 10.7 Eco-efficiency of food commodities.

commodities that were both costly and generated high amounts of waste. Fruits, vegetables, sauces, oils and fats were relatively less costly even though they generated higher amounts of waste. Meat, dairy and eggs generated the least waste and were less costly when compared to high-cost foods such as fish and seafood. This graph can help the restaurant focus and prioritise its food prevention strategy, starting with the high cost–high waste group, followed by the low cost–high waste group, and finally the low cost–low waste group.

Observations of the general procedures and practices outside the kitchen revealed a number of broader factors affecting food waste generation. In buffet operations, food was prepared in advance. The quantity of food to be prepared was based on the reservations made for that meal and estimates of additional customers turning up on the day without any reservation. Accurate prediction of the number of expected customers was crucial in avoiding food surplus. In other words, if food was prepared for the actual number of customers being served, then food waste could be minimised. In order to achieve this, pre-booking is essential. According to the head chef of this case study, in order to ensure sufficient food on the lunchtime buffet, approximately 30 per cent more food was prepared than what was required for the actual bookings. This food surplus also served to satisfy the customers' expectations for variety and 'value for money', as discussed earlier on in this section. This strategy ensured the lunchtime buffet did not run out of food; however, it also contributed to excessive food surplus production, which in turn led to significant buffet leftover food waste.

Many buffet operations have strict policies on the maximum time duration food can be left on the buffet. These policies aim to ensure food safety. The restaurant in this case study had in place a policy specifying that food items should not be left on the buffet for periods longer than four hours. For example, if a dish was served during the breakfast buffet and it was not consumed, it could not be served again at lunchtime and had to be discarded. Although the policy aimed to ensure the food served was fresh and safe for the customers' benefit, it led to significant quantities of buffet leftover waste.

Poor communication and coordination between the different departments in charge of bookings (sales department), food provisioning (purchasing department), food preparation (kitchen), and operations (waiting staff) was also a factor contributing to food waste generation. Clear and timely communication and effective coordination between the departments has the potential to avoid excessive food waste generation. This is especially relevant in instances where changes are made to the initial booking. Effective communication and coordination was sometimes problematic, especially since the different departments had different and often conflicting priorities. The overall mission and values of the departments were the same and in line with company policy. However, when these values were translated into department-specific targets, conflicts became evident. An example of this was apparent within the departmental evaluation system. The sales department was evaluated on the volume and economic value of bookings, the purchasing department on ensuring costs remain low, and the kitchen and operation staff on the quality of service and food. Aligning and connecting these departmental targets back into the company's central values could result in better coordination and effective teamwork between the departments.

Discussion

The discussion focuses on the connection between food waste's biophysical patterns and social practices observed, as presented in the previous section.

More specifically, the discussion explores how these findings relate to the changing food consumption patterns and practices of contemporary Malaysia.

One of the main drivers for food waste generation observed at the hotel restaurant was the restaurant's food safety policy. This policy limited the amount of time food can be left on the buffet and posed the most significant barrier to food surplus recovery and redistribution, even within the operations of the restaurant. A blanket policy such as this removed any decision-making power from the individual in charge; for example, the chef judging whether a food is safe or not based on when it was cooked, its ingredients, the time it has been left on the buffet, among other factors that can influence its properties. This risk adverse policy helps to eliminate the possibility of a food safety breach. This policy resulted in excessive buffet leftover waste that could have been avoided if a procedure was introduced for assessing the freshness of the food and deciding how it should be handled. This food safety policy was 'imported' from the headquarters of the international hotel chain, ensuring all restaurants under the brand name have the same food quality standards. It is an example of a trend observed in Malaysia, whereby international quality standards are perceived by consumers to be superior to local practices.

High aesthetic food standards were another example of the increasingly global norms around what buffets should deliver. Elaborate design in food presentation was a driver for food waste observed in the case study. This practice gave rise to avoidable food waste, especially preparation waste involving fruits and vegetables used as decoration for dishes. On the one hand, the high aesthetic standards represented sophistication, luxury and ultimately superior food quality for the consumer. On the other, the relatively lower price of food produce in Malaysia did not incentivise menu design that encouraged the reuse of edible fruit or vegetable cuttings in other dishes (however, certain foods such as imported products and luxury items are subjected to goods and services taxation as of April 2015, which could eventually promote food waste prevention). This resulted in low eco-efficiencies for these food groups as illustrated in Figure 10.7. The limited economic incentive to improve eco-efficiency, coupled with the growing trend for elaborate food presentation, led to doubly damaging consequences in terms of food waste generation.

Another factor driving food waste generation was the consumers' expectations of a continuously full buffet with excessive amounts of different food items on offer. This excess was exactly what the customer associated with good value for money and satisfaction (as illustrated by the chef's and customers' comments presented in the results section). The fact that quantity and variety, not necessarily quality, of food determined customers' satisfaction highlighted the fact that abundance and extravagance were perceived as synonymous to food of high standards and made up a superior food consumption experience. This superior food experience was perceived as something imported, challenging the traditionally modest local values associated with food consumption, particularly in the context of an international hotel restaurant. Conversely, the restaurant was not willing to challenge customers' expectations in fear of losing business, and actually promoted variety and quantity as a means to attract customers. Promotional material for the restaurant's buffet drew attention to the large number of food items rather than the freshness of ingredients or the culinary skills of the chefs. Why this food consumption trend is emerging and whether it is entirely new are questions requiring further investigation. While these questions remain to be answered, this trend is becoming increasingly popular, even at more affordable restaurants, and leads to avoidable food waste generation.

The examples above demonstrated how imported – or perceived as imported – discourses of food consumption, such as excess, food safety and aesthetic standards, are interpreted and adopted in the Malaysian context, and how they result in excessive food surplus and ultimately waste generation. Whether these increasingly global norms are a response to the changing consumer expectations and perceptions of food quality, or whether consumers' perceptions are influenced by the introduction of imported trends is a debate outside the scope of this study and requires further research. However, this study has shown that the growing popularity of the observed trends has a direct and detrimental effect in relation to food waste generation and at the moment these trends show no signs of slowing down.

Conclusion

The case study presented in this chapter revealed a number of different factors that give rise to food waste within the food service sector. The consumer's expectations, perceptions and social norms converging from around the world are some of the key drivers of food waste generation. Factors related to the nature of the restaurant, for example, the type of service provided, including buffet style, and the fact that food cooked from scratch from fresh ingredients, also lead to food waste generation. The menu design, portioning and high aesthetic requirements for food decoration were also identified as significant factors. Finally, other factors that contribute to food waste relate to the restaurant's organisational structures, policies and procedures. These include the food safety policy requiring food to be discarded after four hours, inefficiencies in the booking system, and finally the lack of communication and coordination between different departments with the restaurant.

These drivers are diverse in nature and require different approaches and different levels of intervention in order to ultimately achieve food waste prevention. Restaurant food waste is an issue that requires a multi-level approach, from top policy level (for example, restructuring the buffet food safety policy, introducing a food surplus redistribution system), to bottom-level interventions dealing with daily operations (for example, training of staff to cut more efficiently, designing menus that allow for the utilisation of all edible parts of food in several dishes), and across different areas and levels of organisations (such as managing consumer expectations, perceptions and practices, improving booking systems, and communication between different departments). Most importantly, this study highlights the need to see food waste as part of food consumption rather than as a separate issue, and as such recognise it as inherently related to social practices and not solely the material context within which it is generated.

Determining the connections between the social and material contexts that give rise to food waste was only possible through an interdisciplinary approach in the research design. Tools from industrial ecology, and approaches from ethnography and grounded theory, were employed to collect and analyse quantitative and qualitative data. Insights on social practices were used to enrich our understanding of the food waste generation process and explain the patterns and relationships that emerged, in order to identify opportunities for food waste prevention. This case study strengthens the argument for interdisciplinary research, especially when investigating complex and contemporary phenomena that are bound by their social and material contexts, like food consumption.

References

Agamuthu, P., H. Fauziah and K. Khidzir. 2009. 'Evolution of solid waste management in Malaysia: Impacts and implications of the Solid Waste Bill, 2007.' *Journal of Material Cycles and Waste Management* 11(2) (June 25): 96–103.

Al-Domi, H., H. Al-Rawajfeh, F. Aboyousif, S. Yaghi, R. Mashal and J. Fakhoury. 2011. 'Determining and addressing food plate waste in a group of students at the University of Jordan.' *Pakistan Journal of Nutrition* 10(9): 871–878.

Beretta, C., F. Stoessel, U. Baier and S. Hellweg. 2013. 'Quantifying food losses and the potential for reduction in Switzerland.' *Waste Management* 33(3): 764–773.

Betz, A., J. Buchli, C. Göbel and C. Müller. 2015. 'Food waste in the Swiss food service industry – magnitude and potential for reduction.' *Waste Management* 35: 218–226.

Bostok, M. 2014. 'Sankey diagrams software.' ramblings.mcpher.com.

Evans, D. 2011. 'Beyond the throwaway society: Ordinary domestic practice and a sociological approach to household food waste.' *Sociology* 46(1) (October 27): 41– 56.

Evans, D., H. Campbell and A. Murcott. 2013. *Waste Matters: New Perspectives on Food and Society.* Edited by D. Evans, H. Campbell and A. Murcott, 1st edn. Oxford: Wiley-Blackwell/The Sociological Review.

FAO. 2013. *Food Wastage Footprint. Impacts on Natural Resources.* Rome: FAO.

FAO. 2014. *Global Initiative on Food Losses and Waste Reduction.* Rome: FAO.

Garnett, T. and A. Wilkes. 2014. *Appetite for Change: Social, Economic and Environmental Transformations in China's Food System.* 94: 4-5. Oxford: Food Climate Research Centre. www.fcrn.org.uk/sites/default/files/fcrn_china_mapping_study_final_pdf_2014.pdf (accessed 8 June 2016).

Glaser, B. and A. Strauss. 1967. *The Discovery of Grounded Theory.* Chicago: Aldine.

Gustavsson, J., C. Cederberg, U. Sonesson, R. van Otterdijk and A. Meybeck. 2011. *Global Food Losses and Food Waste: Extent, Causes and Prevention.* Rome: FAO.

Johari, A., S. Ahmed, H. Hashim, H. Alkali and M. Ramli. 2012. 'Economic and environmental benefits of landfill gas from municipal solid waste in Malaysia.' *Renewable and Sustainable Energy Reviews* 16(5) (June): 2907–2912.

Jupp, V. 2006. *The SAGE Dictionary of Social Research Methods*, pp. 179–180. London: Sage Publications.

Malaysian–German Chamber of Commerce. 2012. *Market Watch 2012. The Malaysian Food Industry*. Kuala Lumpur: Malaysian–German Chamber of Commerce.

Padfield, R., E. Papargyropoulou and C. Preece. 2012. 'A preliminary assessment of greenhouse gas emission trends in the production and consumption of food in Malaysia.' *International Journal of Technology* 3(1): 56–66.

Papargyropoulou, E., R. Lozano, J. Steinberger, N. Wright and Z. Ujang. 2014. 'The food waste hierarchy as a framework for the management of food surplus and food waste.' *Journal of Cleaner Production* 76 (August): 106–115.

Papargyropoulou, E., N. Wright, R. Lozano, J. Steinberger, R. Padfield and Z. Ujang. 2016. 'Conceptual framework for the study of food waste generation and prevention in the hospitality sector.' *Waste Management* 49 (March): 326–336.

Pemandu (Performance, Management and Delivery Unit). 2012. 'Developing Greater Kuala Lumpur/Klang Valley as an engine of economic growth.' In *Economic Transformation Programme: A Roadmap For Malaysia*, pp. 123–162. Kuala Lumpur: Government of Malaysia.

Quested, T., D. Parry, S. Easteal and R. Swannell. 2011. 'Food and drink waste from households in the UK.' *Nutrition Bulletin* 36(4) (December 8): 460–467.

Saunders, M., P. Lewis and A. Thornhill. 2009. *Research Methods for Business Students*, 5th edn. Harlow: Pearson Education.

Waste and Resources Action Programme. 2007a. *Understanding Food Waste. Research Summary*, 7–8. Banbury: Waste and Resources Action Programme.

Waste and Resources Action Programme. 2007b. *We Don't Waste Food! A Householder Survey*, 4–5. Banbury: Waste and Resources Action Programme.

11 Convenient food, inconvenient waste

Systems of provision meet social practices in Seoul

Keith Lee

We discard an estimated one-third of all food produced for human consumption, with significant social, environmental and economic consequences (Gustavsson et al. 2011). Food waste threatens global food security, and reducing food waste is vital to addressing malnutrition (Godfray et al. 2010). Food waste also contributes to climate change: it produces methane in landfills, and agriculture is a large source of greenhouse gas emissions. Venkat (2012) estimated that in 2009, avoidable food waste in the United States of America was worth USD 200 billion and accounted for 113 million tonnes of carbon dioxide equivalent per year, or 2 percent of net national greenhouse gas emissions. Sixty-nine percent occurred in production and processing, 6 percent in packing, 12 percent in distribution and retail, and 14 percent from disposal. Rapid urbanization in the developing world will exacerbate these costs (Adhikari et al. 2009), and inadequate waste infrastructure is expected to worsen local pollution and threaten public health (Wilson et al. 2012). To avoid these impacts, further research is required to understand the drivers of food waste and to support improved management and policies.

Research has not only suggested that age, household size and certain attitudes are possible drivers of household food waste (Koivupuro et al. 2012; WRAP 2013), but has used practice theory to show how the disruption of grocery shopping and meal-planning practices in the course of everyday life generates food waste (Quested et al. 2013; Evans 2014). Although studies have touched upon how household food waste is linked to systems of provision such as the food retail system, few have studied the emergence of these systems and how they influence household behavior. This approach is important given the connections between urbanization and systems of provision and their capacity to shape and constrain daily life. This chapter will extend the application of practice theory to household food waste by exploring its causes in relation to food retail modernization in Seoul, South Korea (henceforth 'Korea').

I begin with my conceptual framework, which uses practice theory to explore the relevance of time-use and systems of provision for understanding everyday behaviors that lead to household food waste. I then discuss my research methodology, based on interviews and Korean Time Use Survey data, as well as the research context, where I explain the food waste situation and dynamics of food retailing in Seoul. Next, I present and discuss my findings, exploring changes in food practices and their implications for household food waste. I conclude the chapter with policy recommendations and suggestions for future research.

Conceptual framework

Practice theory has emerged as a theoretical perspective from which to examine household consumption and sustainability issues, among them food waste. Practice theory focuses on the practices that comprise everyday life. Here, a practice refers to 'a routinized type of behavior' (Reckwitz 2002: 249) that is recognizable by society in the abstract – cooking or reading, for example – and consequently persists over time. Practices are the building blocks of social order, and simple practices can form more complex ones. For example, cooking is comprised of simpler practices such as cutting, frying, measuring and timing. How practices are performed can change over time when individuals introduce variation via adaptation, improvisation and experimentation; such efforts could result from personal expression and experience or from the influence of social, institutional and technological contexts (Warde 2005).

Practice theory conceptualizes consumption and waste as embedded in and linked by practices. Prior accounts define consumption as the outcome of individual efforts to maximize utility, establish self-identity, or demarcate social status (as discussed in Halkier 2010 and Røpke 2009). Warde (2005) however, argues that consumption is not a practice itself, but exists to enable practices by allowing people to access the goods and services required to perform these practices. All practices hence involve consumption in one form or another. Others have suggested that waste is the outcome of a profligate culture of consumption (Strasser 2000), but Evans (2012) finds that individuals often feel guilty about what they throw away and that waste cannot be considered separately from how people buy, store, manage, prepare and eat their food, as well as how they coordinate these practices in daily life. Consequently, waste should instead be considered as emergent from everyday life.

Coordinating practices can be tricky considering they take up time, a finite resource. Performing one practice leaves less time for others. Therefore, accounting for time management is vital to understanding individual behavior. This is particularly relevant in modern society as the notion of time-scarcity, or the perception of not having enough time, grows with income (Hamermesh and Lee 2007). People develop particular expectations and understandings regarding the when, how long and sequence of particular practices, leading to the normalization of practices' temporal characteristics. For example, a three-course 'dinner' is often understood to begin with an appetizer and to finish with dessert (Shove 2009). Practices may require 'synchronization and coordination' with other practices or individuals by requiring one to be in certain places at specific times – family holidays, for example (Southerton 2006: 443). Temporality and practices shape each other simultaneously; practices influence temporality through particular requirements (for example, dinner parties are understood to be evening events) and temporality influences practices through the existence of 'collective and personal rhythms' (for example, the daily cycle of work commutes) (Southerton 2012: 344).

Culture can affect the temporality of practices by influencing how people understand and organize time. Different cultures exhibit varying tendencies towards polychronicity, the tendency to perform multiple practices simultaneously. Polychronic cultures tend to place less importance on adhering to plans as they have a less linear view of time (Bluedorn 2002), reducing the importance of sequencing and punctuality. Culture also influences pace of life (Brislin and Kim 2003), with the implication that practices are likely to be performed at different speeds depending on cultural context. These factors suggest that the coordination of daily life varies greatly across cultures, as does the extent to which people adhere to or deviate from their daily schedules.

The temporality of practices in modern society has been altered by convenience, which has allowed people to compress and re-order time to fit their needs (Warde et al. 1998). For example, freezers and microwaves provide shortcuts to meal preparation by saving labor and cutting cooking time (Shove and Southerton 2000). However, greater convenience has led to increasingly fragmented individual schedules; whereas they were previously governed by collective temporal markers like the day of the week, increased convenience has enabled individuals to be more independent from 'formalised, collectively shared, temporal structures' as they can now shift or double-up certain practices (Shove 2003: 412). This increased complexity paradoxically creates an even greater need for convenience so that individuals can maintain coordination of not only their own practices, but those that involve others too. Shove (2003) identifies growing reliance on convenience as a potential problem if it creates convergence towards, and lock-in of, undesirably resource-intensive norms of daily life.

One concept that illuminates the relationships between consumption, waste and convenience is that of systems of provision (SOP), which was originally introduced to emphasize a technological-historical perspective on consumption patterns (Fine and Leopold 2002). SOPs most readily refer to infrastructure such as power and gas, sewage and transportation systems, but have also been conceptualized as the physical and institutional infrastructures that enable access to goods and services (Southerton et al. 2005), providing the structure within which people carry out practices in daily life. In the food context, SOPs include the food production and distribution system, namely, food production, harvest, storage, distribution and retail. SOPs also exist at smaller scales. For instance, most household kitchens contain a stove, refrigerator and miscellaneous items such as a toaster and microwave among others, forming an 'infrastructure of the home' (Shove and Walker 2010: 473). These are all dependent upon larger systems like the power and gas network and are also delivered by their own SOP (for example, manufacturing, retail), revealing the nested and multi-scalar nature of SOPs.

SOPs can enable, constrain or contextualize consumer behavior (Spaargaren 2003, cited in Shove and Walker 2010), which I argue takes place through the provision of convenience – many SOPs improve individuals' time-management abilities, either through saving labor or altering mobility, thereby influencing how individuals coordinate their practices in time and space. SOPs' influence on practices is important for understanding household food waste, given the temporal characteristics associated with food and its related practices. For example, from the moment consumers buy fresh ingredients, they enter a race against time to use them before they expire. Refrigeration allows consumers to, in a sense, manipulate time by slowing down the decay of perishables, thereby forestalling the race and ensuring a ready supply of food (Shove and Southerton 2000). However, refrigeration also allows busy consumers to stockpile food, sometimes to the point whereby they no longer have enough time to consume their food before it goes bad or starts causing concerns about food safety (Graham-Rowe et al. 2014).

The well-stocked refrigerator, together with home cooking, has also become symbolic of good housekeeping (Graham-Rowe et al. 2014). However, the well-stocked refrigerator means that even when consumers do their best to avoid food waste by planning their grocery shopping and meal routines, everyday life's contingencies, such as an unexpected business trip or a call from friends, can incite abrupt schedule changes, preventing consumers from carrying out their plans and leaving food to run its course towards abandonment (Evans 2012).

SOPs can provide convenience and influence food waste by modifying when and where consumers do their grocery shopping. At the most basic level, supermarkets permit one-stop shopping, reducing grocery shopping frequency while increasing the quantities bought per trip (Shove and Southerton 2000). This potentially contributes to household food waste by making perishable inventories larger and harder to manage (Williams et al. 2012). The reality is likely more complex. Depending on surrounding transportation networks (another SOP) and the location of supermarkets, access to food can be either convenient or inconvenient. Other retail factors also intervene, such as opening hours, food packaging and portions, shelf-life and labeled expiration dates, as well as store characteristics including product placement, promotions, smells or even music. These SOP-related factors, together with household-level factors such as life-stage, habits, preferences, diet and schedules, underlie different patterns of grocery shopping and, in turn, differentiated food wastage patterns.

Food practices and their place in the coordination of daily life have converged towards a pattern whereby household food waste becomes almost a natural byproduct of domestic life. As shown above, SOPs retain a role in the dynamics of household food wastage, underlining the importance of continued research that leverages this aspect in relation to practice theory. Yet current research on how SOPs influence everyday practices is not only limited, but also very much focused on the Western context. Given the growing waste issues faced by developing cities around the world, we need to examine different institutional, technological and sociocultural contexts, to determine how and in what way interactions between SOPs and domestic food practices engender different modes of everyday life, how these are changing over time, and what this means for household consumption and food waste.

Methodology

The remainder of this chapter chiefly draws from my analysis of Korean Time Use Survey (KTUS) data and fieldwork I conducted in Seoul in July and August 2013 and from August 2014 through July 2015. The fieldwork involved observations made during trips to different food retailers and of daily life in Seoul, as well as interviews with academics, government officials and thirteen households.[1] Household interviews were conducted with the person primarily in charge of grocery shopping and cooking at home, and were semi-structured, using pre-prepared questions about the household's food-related routines as the basis for conversation.

Research context

As part of the broader municipal solid waste problem, food waste has attracted Korean policymakers' attentions for decades. Recognizing the limitations of landfills imposed by the country's small size, the government banned food waste from landfills in 2005 and introduced volume-based pricing for food waste nationwide in 2013, under which households and small businesses are required to pay a per-liter food waste disposal fee (Kim and Kim 2012). The volume of food waste generated per capita in Seoul peaked in 2009 and has slowly declined to around 300 g per capita in 2013 (Figure 11.1), 70 percent of which comes from homes and small restaurants (Ministry of Environment 2014).[2]

Despite the apparent success of Korea's waste policy, there are several caveats. First, enforcing compulsory waste segregation is tricky, and it is unclear

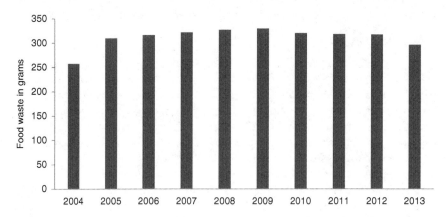

Figure 11.1 Per capita per day food waste in Seoul, 2004–2013.
Seoul Metropolitan Government.

how much reduction has been attributable to illegal dumping or reductions in moisture content. Households can also become desensitized to volume-pricing over time without sustained price increases, which are politically unpopular (K.-S. Park,[3] personal communication). Given these limitations and the importance of developing policy that goes beyond targeting individuals (Shove 2010), it is worthwhile to understand how system-level influences like SOPs may be linked to food waste in order to identify further policy options for managing food waste (Evans 2014).

In Europe and North America, supermarkets and hypermarkets[4] are often taken for granted as the default source of groceries. Elsewhere however, the rise of supermarkets over more traditional food retail has been compressed over a shorter period of time (Reardon et al. 2003), accentuating its effects and creating a more diverse food retail landscape. Known as 'retail modernization' or 'supermarketization', this process is associated with economic development and the adoption of Western consumerist lifestyles. Korea is one country where this has taken place – prior to retail sector liberalization between 1989 and 1996, there were no supermarket chains, and food retailers were mainly comprised of family-run stores and traditional market sellers (Cho et al. 2013). Liberalization was a government-led initiative to improve productivity and efficiency in the retail and distribution industries (Suh and Howard 2009); it allowed the entrance of foreign retailers like Carrefour, Wal-Mart and Tesco, which introduced the novel hypermarket format (Jin and Kim 2003). While urbanization, rising disposable incomes and female labor force participation increased sensations of time-scarcity and raised consumers' opportunity cost of time (Hamermesh and Lee 2007), hypermarkets provided convenient one-stop shopping in pleasant and modern shopping environments, a factor contributing to their rapid expansion and growth in popularity (Figure 11.2).

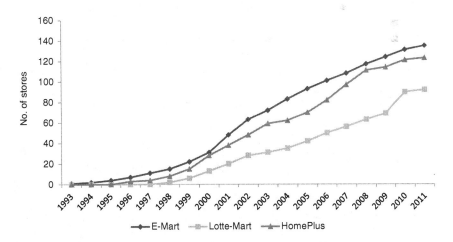

Figure 11.2 Hypermarket growth (includes stores acquired from other chains), 1993–2011.
Adapted from Cho et al. 2013.

However, all the foreign retailers who entered Korea in the 1990s have since withdrawn, mostly reflecting their failure to accommodate Korean consumer preferences. Their low-pricing strategies and spartan warehouse-style stores repelled brand-loyal and quality-conscious locals (Kim 2008), who prefer attractive product displays, lively sales promotion and customer service (Gandolfi and Strach 2009). Korean cuisine is vegetable-heavy, contributing to a cultural preference for freshness that underlines a tendency to shop frequently and buy in small quantities. Western retailers failed to account for this preference (Coe and Lee 2006) and compounded this mistake by locating in areas not easily accessible by foot (Kim 2008). The case of Homeplus throws these particularities further into relief. Originally owned by Tesco,[5] Homeplus successfully localized by placing fresh food displays near store entrances, developing private brands and cultivating local supplier relationships (Coe and Lee 2006). Homeplus also pushed the boundaries of one-stop shopping by providing other amenities, including food courts, art galleries, wine bars and health clubs (Coe and Lee 2013). In short, the hypermarket format has taken on a distinctly local flavor in Korea.

Korea's retail modernization has also involved chain supermarkets and convenience stores. Following their hypermarket successes, the local 'big three' of Emart, Homeplus and Lotte Mart began opening supermarkets in the mid-2000s. Dubbed 'SSMs' (super-supermarkets), these are sized between hypermarkets and family-run stores.[6] SSMs sell food and household products, employing a strategy that balances product selection, prices, promotions and customer service (Moon 2010). The number of SSMs in Korea grew from 273 in 2005 to 699 in 2009. Finally, chain convenience stores are the latest trend in the Korean grocery retail sector, and are distinguished by their small size[7] and extended opening hours. Although previously selling sundries and ready-to-eat food, convenience stores now increasingly stock fresh foods like eggs, vegetables and fruit. As with Korea's hypermarkets and supermarkets, convenience stores are predominantly owned or franchised by corporate retailers.

By 2014, hypermarkets, supermarkets and convenience stores accounted for 73 percent of grocery retail revenues, and traditional market sellers and family-run stores (collectively 'traditional retailers') held the remainder (Euromonitor 2014). Traditional retailers' shrinking market share and the ensuing backlash against corporate retailers (Choe 2012) has prompted government regulation of corporate retailers, along with investments to upgrade traditional market facilities. For example, hypermarkets and SSMs in Seoul must now close two days every month and may not operate between midnight and 8am (traditional markets tend to close around 8pm while family-run stores sometimes close as late as midnight). Hypermarkets and SSMs have also been barred from opening within 1 km of traditional market areas (Distribution Development Industry Act 2012; Song 2012). Although these regulations have helped sales at traditional markets (Kim and Hallsworth 2015), family-run stores have been converting their shops into franchised convenience stores in order to survive (Cho et al. 2013). In response, corporate retailers have begun moving into the less-regulated convenience store segment, prompting accusations of opening 'mini-SSMs' in the guise of convenience shops in order to bypass regulations (S. Kim 2013). Internet grocery retailing is also expected to become increasingly important (Euromonitor 2013). Though the eventual outcome of food retail competition in Korea's urban areas remains uncertain, the impact of corporate retailers has been significant.

By describing recent changes in the Korean grocery retail sector, I have sketched the distinctive nature of the SOPs that undergird Korean households' food-related practices. The question that follows is what this ongoing Korean flavor of retail modernization has meant for food-related practices among Seoul's households, and what this entails for food waste. It is also relevant to ask how SOPs are responding to or shaping food-related practices, and how everyday life might simultaneously shape and be shaped by SOPs. In the next section, I will use KTUS data to assess trends in the temporal and demographic dimensions of food-related practices, link these changes to Korea's retail modernization, and finally discuss the implications for food waste.

Research results

In order to understand changes in food-related practices over time, I examined KTUS data between 1999 and 2014. First, I attempted to assess the possibility that Korean shoppers are increasingly shopping less often (for example, once a week versus two to three times a week) and buying more per trip. This could be reflected in a shift over time in the percentage of people shopping for groceries on weekdays to the percentage shopping on weekends and for longer durations. Figure 11.3 shows the increase in the percentage of people shopping on weekends over most times of the day.

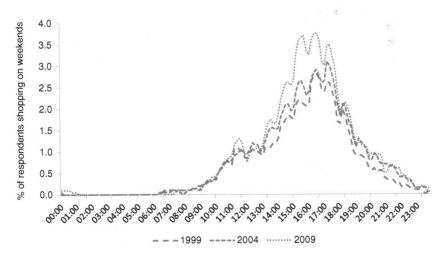

Figure 11.3 Percentage of respondents (aged 10+) who reported spending time shopping for household-related purposes at different times on weekends.
Korea Time Use Survey 1999, 2004, 2009.

However, an increased tendency towards once-a-week weekend shopping trips would be accompanied by a decreased percentage of people shopping on weekdays, which is not supported by the data (Figure 11.4). Conversely, this percentage has increased marginally over the period shown. This suggests that Korean consumers are still shopping several times during the week while adding weekend trips, or that any trend towards once-a-week weekend shopping among some has been confounded by an increasing tendency among others to shop on weekdays.

Examining disaggregated data provides further insights. Figure 11.5 compares weekday shopping patterns for unmarried and married shoppers, revealing minor changes in the percentage of married people shopping during the week and an increase in the percentage of unmarried people shopping on weekdays. This suggests that married shoppers are not giving up on weekday shopping trips, while unmarried shoppers are increasingly going to stores on weekdays.

The increase in shopping taking place after 8pm is noteworthy, as it reflects how Koreans are increasingly utilizing the modern retailers' longer opening hours. Unlike traditional markets, which rarely stay open past 8 or 9pm, hypermarkets and SSMs usually stay open until midnight, and convenience stores are open twenty-four hours. These extended hours are crucial in light of how Koreans work some of the longest hours in the world (Lee et al. 2007) and face ever longer commutes (Jun et al. 2013). Office culture also often demands that workers socialize over drinks in the evenings, even during the week, meaning workers may not have time to buy groceries until late after work. Modern grocery retailers' extended hours ease the temporal restrictions on when consumers do their grocery shopping, making it easier for them to fit the

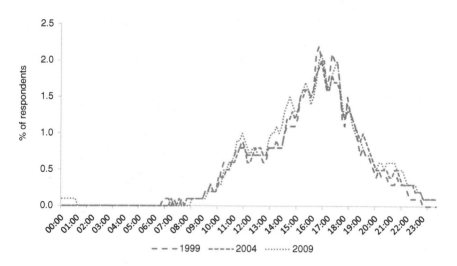

Figure 11.4 Percentage of respondents (aged 10+) who reported spending time shopping for household-related purposes on weekdays.
Korea Time Use Survey 1999, 2004, 2009.

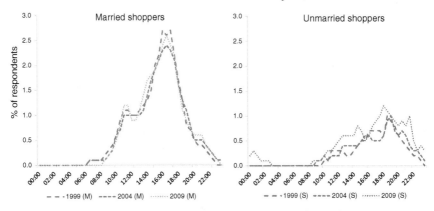

Figure 11.5 Percentage of married (M) and single (S) respondents (aged 20+) who spent time shopping for household-related purposes on weekdays.
Korea Time Use Survey 1999, 2004, 2009.

practice of grocery shopping into their everyday schedules. Extended opening hours may also have unanticipated effects on shopping practices. Two fifty-something housewives I interviewed, Mi-young and Young-sook,[8] described how they not only shopped several times during the week, but also at multiple venues for the ingredients deemed best at each location. However, they also enjoyed shopping in the evenings after dinner. This was not due to any time constraints, but because the evening clearance discounts offered by retailers offered the thrill of bargain-hunting. These observations and the above data suggest how the temporal characteristics of SOPs can shape food-related practices in expected and unexpected ways.

Besides extending their opening hours, modern grocery retailers have also adapted to Korean workers' long working hours and busy lifestyles in other ways. In particular, Homeplus drew international attention by experimenting with virtual grocery stores, which involved pasting displays of shelved grocery items in Seoul's subway stations and bus stops. This allowed commuters the chance to purchase needed items with their mobile phones while waiting for their bus or train. Items purchased this way before 1pm would be delivered the same evening ('Tesco opens world's first virtual store' 2011). In a sense, this innovation is an outgrowth of the broader culture of home delivery that pervades the Korean grocery sector. Given a minimum amount spent, free home delivery is widely available from physical store locations, not only from hypermarket and SSM chains, but also from family-run grocery stores. This allows Korean consumers to walk to the grocery store while freeing them from having to carry their groceries home. Such delivery services and innovations, such as Homeplus' virtual stores, together with extensive online shopping options and high overall walkability mean that Korean consumers benefit from unprecedented convenience and access when it comes to grocery shopping.

Concurrent with the rise in percentages of unmarried shoppers on weekdays the percentages of married and unmarried people shopping on weekends have increased (Figure 11.6). Interestingly, the percentage of married men shopping on weekends grew from 12 percent in 1999 to 21 percent in 2014. Since weekday shopping trips have not declined, this increase suggests that the additional weekend shopping trips may represent opportunities for leisure and/or family bonding rather than reflect any transition towards once-a-week grocery shopping. With the efforts made by hypermarket chains such as Homeplus to increase the array of services and products offered in their stores, traditional retailers struggle to compete with hypermarkets in terms of providing an integrated venue for family leisure. As early as 2003, 46 percent of Korean hypermarket shoppers were found to have been motivated by leisure or socialization (Jin and Kim 2003), and the data in Figure 11.6 may reflect this.

Interestingly, there has also been an increase in the overall proportion of unmarried people who indicated shopping for groceries at all (Figures 11.5 and 11.6). This may be related to growth in the number of single-person households, which has increased from 9 percent to 25 percent of the population between 1990 and 2010 as a result of urbanization, later marriage, low fertility and rising divorce rates (Kim 2014). There has also been growth in the percentages of unmarried men and women who spent time preparing food at home from 15 percent and 29 percent in 1999 to 22 percent and 32 percent in 2009, respectively. These trends have been attributed to the recent popularity of cooking shows (Gallup Korea 2015). Single-person households are an attractive customer group for retailers, and as such these trends are likely to drive adaptive changes in modern retailers' operations and promotional campaigns.

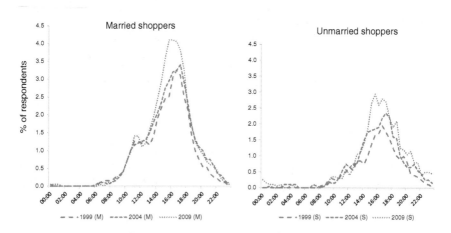

Figure 11.6 Percentage of married (M) and single (S) respondents (aged 20+) who spent time shopping for household-related purposes at different times on weekends.

Korea Time Use Survey 1999, 2004, 2009.

Shrinking Korean households and the growing tendency among unmarried consumers to cook at home increases the tendency for packaging sizes to exceed consumers' needs. When consumers are forced to buy food in greater quantities than needed, the result is often the spoilage of partially used ingredients – one way Western supermarkets are blamed for food waste (Graham-Rowe et al. 2014). In the Korean case, however, these trends have been noted by modern retailers, who have expanded their offerings of ready-to-eat convenience foods and food packaged into small portions, particularly spoilage-prone fruits and vegetables (B. Kim 2013). As such, it is common to see half an onion sealed in a vacuum pack for sale in convenience stores, SSMs and hypermarkets alike. This response not only reflects how SOPs are mutable and responsive to broader change, but also illustrates the connections between small and large-scale SOPs. Selling food in smaller portions addresses not only the smaller quantities consumed by single-person households, but also the smaller scale of these households' kitchen infrastructure – smaller refrigerators and cupboards provide limited storage space, while cramped kitchens with small surfaces often make elaborate food preparation more trouble than it is worth.

The 2014 KTUS data also revealed the extreme gender division of labor in Korean households. Just 20 percent and 15 percent of unmarried and married men, respectively, reported spending any time preparing food. In contrast, whereas 27 percent of unmarried women reported spending any time preparing food, the equivalent figure for married women was 88 percent. The gap was no smaller among dual-income couples, among whom the figures were 88 percent for wives and 13 percent for husbands. Motherhood further increased the extent to which married women prepared food at home regardless of employment status – a prior study based on 1999–2009 KTUS data found maternal employment actually increased the probability that working mothers spent time on food preparation, albeit for shorter durations than non-working mothers (Chang and Lee 2012).

These data reflect several characteristics of Korean domestic life that contribute to time stress experienced by Korean women. Korean men have traditionally not helped out with household chores as they were historically forbidden from entering the kitchen (Lee 2005). Korean women thus face heavy expectations to cook for their families (Bak 2014), regardless of employment status. The time-stress these expectations create is potentially exacerbated by the characteristics of the Korean diet. Unlike other Asian countries that experienced rapid economic development, Korea's dietary transition towards increased fat and carbohydrate intake has been very slow; instead, vegetable consumption has remained high, reflecting an adherence to the traditionally low-fat and vegetable-heavy Korean diet (Lee et al. 2012). This has been attributed to Koreans' conservative food attitudes, government policies and government-initiated movements that promoted traditional Korean cuisine in order to address dietary health concerns and ensure continued support for local vegetable production. These efforts involved cooking training for housewives, nutritional education and mass media campaigns (Lee et al. 2002). However, preparing vegetables is highly time and labor-intensive due to the need for extensive trimming and peeling by hand. Korean cuisine is also notable for its wide array of side dishes, or *banchan*, which vary from region to region and are traditionally meant to harmoniously complement each other when served (Oum 2005). The wide range of ingredients and knowledge required consequently make Korean cooking a time-intensive endeavor.

The nature of Korean cooking and the expectations faced by Korean women illustrate how food-related practices 'outcompete' other practices for places in their daily schedules. Prior research suggests that in terms of food choices, people respond to increased time scarcity in several ways, including eating faster, limiting mealtimes and using time-saving substitutes (Jabs and Devine 2006). The latter might include eating out, ordering in or buying convenience foods. The 2010 Korean Family Survey (KFS) of 2063 households found that 65 percent and 62 percent of couples with children and married couples, respectively, cooked homemade meals at least once a day. Twenty-eight percent of KFS respondents identified cooking duration as the biggest challenge to cooking at home, but only 11 percent of respondents listed not having enough time as the main reason for eating out. These data suggest that Korean households are persisting with home-cooked meals. However, working wives spent twenty-one fewer minutes on average than housewives on food preparation (KTUS 2014), which suggests they are taking advantage of time-saving devices or preparing simpler meals when cooking at home.

Among the time-saving opportunities are the new forms of food provided by modern retailers: hypermarkets and SSMs increasingly sell fruits and vegetables in pre-washed, pre-trimmed, pre-peeled or pre-sliced packages. Although the concept is not new to Western supermarkets, a glance at the 'convenience vegetables' for sale reveals the extent to which Korea's modern grocery retailers have taken this concept (Figure 11.7). For example, beyond

Figure 11.7 'Convenience vegetables' and prepackaged salads on sale at Homeplus.

pre-washed salad mixes, one can find pre-diced sweet potato and pumpkins, pre-shredded and pre-seasoned spring onions and a pre-diced vegetable mix specifically for use in Korean omelets. In the fresh meat and seafood sections, one may purchase trays of ingredients in the right proportions for use in stews. Fresh ready-to-eat foods on sale are also not lacking in variety, and there is a dizzying variety of pickled and fermented vegetables. Rather than being protected behind glass counters, these are heaped on large open trays, emulating the displays of traditional markets.

Besides modern retailers, relatives form another SOP that provides significant convenience to single-person households and married Korean women. It is common, for mothers and mothers-in-law especially, to continue cooking and giving food to married children who live separately, often because they are too busy to prepare food for themselves. According to Bak (2014), this act helps Korean mothers maintain a sense of connection to their children. This is not only gratifying, but may also serve as a source of power. Bak (2014: 26) provides a case in point with her analysis of *kimchi* and social relations:

> For younger married couples, when both side's mothers are willing to provide *kimchi*, things can become competitive between the two mothers. One male college professor confessed that he and his wife always have too much *kimchi* in the refrigerator. To solve this problem, they give the excess *kimchi* to friends. They simply could not tell either of the mothers to stop giving them *kimchi*.

Many of my interviewees told similar stories, although the motherly menus expanded well beyond *kimchi*. Most live alone in Seoul and receive food every few months, either via delivery services or a visit. The standard care package for Seo-yoon, a working professional who lives on her own includes pre-prepared *banchan* and fresh ingredients like onions, sweet potatoes, potatoes, spring onions, cucumbers and zucchinis. Ji-min, a university student, normally receives a month's supply of *banchan* in addition to frozen meat. Such examples reveal the key role that families play in supporting or substituting food practices among Korean households; as household sizes continue to shrink, these alternate systems of provision are only likely to grow in importance.

Discussion

As the reader will recall, prior research has suggested that supermarkets and refrigeration led to a convergence in grocery shopping patterns towards high-volume, once-a-week trips (Watson 2012), and that this may lead to increased food waste due to the challenges of storing large quantities of perishable food (Williams et al. 2012). Examining KTUS data in conjunction with my interview and observational data suggests that even if true in the West, this may be an oversimplification of Korean consumer lifestyles. Albeit a small and non-representative sample, none of my thirteen interviewees had adopted such a pattern, nor did any indicate only ever patronizing one food retailer. It instead appears that Korean consumers may utilize a range of different retailers when shopping for different foods, and at varying frequencies. This illustrates the significance of the diverse Korean food retail landscape. Furthermore, the role of related SOPs, such as the transportation system, is highlighted not only with the example of Homeplus' virtual grocery stores, but also when considering how almost all my interviewees walked or used public transportation and shopped close to home or on their way home from work. In this way, Korean consumers' cultural tendency to shop frequently (Coe and Lee 2006) is supported by urban development patterns that feature high physical accessibility to food. However, without being able to compare quantitative food waste data across different shopping patterns within the same sociocultural context, it remains to be seen whether higher-frequency shopping patterns are associated with less food waste.

Although Korea's retail modernization may not have sparked convergence in shopping patterns, the emergence of hypermarkets in particular has encouraged people to use grocery shopping as an opportunity for leisure. Here, the implications for food waste lie in the possibility that spending more time in a store with leisure as the primary mindset could increase the risk that consumers buy more than needed. In particular, consumers who are grocery shopping for pleasure may have less concrete purchasing goals before entering the store, which may promote impulse purchases (Bell et al. 2011). Furthermore, food shopping contexts such as store layout, shelf arrangements, music, lighting and smells can subconsciously encourage excessive buying (Cohen and Babey 2012). Finally, the potential increase in evening bargain-hunting trips as described by my interviewees may be linked to increased household food waste, if the shorter shelf life of food purchased on discount is not taken into account when planning meals at home.

At first glance, the convenience offered by retailers who appeal to single-person households by selling small-portion vegetables has the potential to help reduce food waste by ensuring that people can buy ingredients in the quantities needed, as well as aiding portioning with pre-measured amounts. However, by replacing naturally protective outer layers with plastic vacuum packaging, as is done for commonly pre-trimmed vegetables like garlic, onions and potatoes, among others, it is possible that the shelf life of these vegetables is reduced, especially if the packaging is subsequently opened but the ingredients are not used all at once. Even if packaged fresh ingredients are equally effective at inhibiting spoilage, the presence of packaging and sell-by dates could create concerns about food safety, leading to their premature disposal if not used in time. It is interesting to note that most of my interviewees who lived alone[9] said they preferred not to buy small-portion convenience vegetables due to higher prices and a perceived lack of freshness. Further research is required to understand which consumer segments are buying such products and how they are being used in food practices, but their potential lack of popularity and use of sell-by dates suggests that when unsold they may also contribute to increased food waste in stores. The environmental impacts from the manufacture and disposal of additional packaging also need to be considered, upstream from consumers.

My interviews also revealed how Korean households supplemented their food provisioning practices via alternate SOPs by receiving food from parents. The potential for increased food waste is clear with Bak's (2014) reference to excess *kimchi*. Although Bak's interviewees mentioned giving away the excess, many of my interviewees ended up throwing away some of this food provided by their mothers. Reasons cited included growing tired of repeatedly eating the same thing, not liking their mothers' cooking, insufficient storage space, not having enough time to eat at home or that there was just too much food to eat. Ji-ae, a bank teller in her thirties, ultimately asked her mother to stop giving her food because she had just resumed work after giving birth, and neither she nor her husband had enough time to eat everything. Although Ji-ae was able to prevent potentially large amounts of food waste through communication with her mother, this aspect of Korean society illustrates how social relations are embedded in a household's food-related SOPs, making it difficult for recipient households to control the quantity and timing of these inflows of food. In households where working mothers continue to shop for groceries, this dynamic complicates their coordination of daily food practices in terms of how much to buy and what to cook, with substantial implications for food waste.

As with the case of refrigeration, the trends in food retail SOPs and parental SOPs alike provide additional convenience while also generating the potential to increase household food waste. Modern retailers have already played a significant role in enabling Korean consumers to maintain their performance of shopping and home-cooking practices via the provision of convenience; they will likely find similar ways to tap into the recent popularity of home cooking. The end result is likely to be increased physical and temporal access to food among a wider variety of consumers, and an increase in the extent to which food practices are imbued, or even dominated by, meanings of leisure. This could promote greater ambitiousness around home cooking while also increasing the tendency to treat food as a commodity input into leisure cooking.

Just as Evans (2012) conceptualizes food waste as the inevitable outcome of competent domestic practice, it is important to consider whether food waste will also become an inevitable outcome of competent food practices that incorporate elements of leisure. Ji-ae provided a possible portent when describing how she would buy ingredients to garnish her dishes for aesthetic reasons, but ended up discarding the remainders because she had no use for them. Among consumers who are relatively new to the kitchen, increased convenience in accessing food and the knowledge of how to prepare it may need to be balanced by increased competence and regularity in managing and preparing food in order to avoid increased food waste.

Conclusion

In this chapter, I have described how, through retail modernization, SOPs in Korea are linked to changes in the nature of food practices. This may take place through the provision of convenience, either via temporal access to food or by providing shortcuts in food preparation. I have linked trends in food-related practices to changing household structure, a wave of popularity in home cooking, the potential for modern retailers to support these trends, and the increased meanings of leisure associated with food practices. While increased household food waste is not a foregone conclusion, it may come about if consumers buy food as part of a trend without engaging with a separate set of food management competencies to avoid food waste.

More broadly, the case of food retail modernization in Korea shows how SOPs evolve over time in response to the social, economic, cultural and institutional contexts of food-related practices, while simultaneously holding the power to structure food-related practices in both expected and unexpected ways. Unlike how particular norms have emerged for other practices such as bathing and indoor temperature control (Shove 2003), the interaction between SOPs and food practices may create more complicated outcomes in terms of how individuals coordinate their daily lives around household provisioning and sustenance. These outcomes are also inextricable from the deeply cultural and symbolic nature of food and food consumption practices, as further demonstrated in several chapters in this volume. In turn, these dynamics have the potential to both alleviate and exacerbate the generation of household food waste, echoing prior work that has highlighted the complexities of household food waste (Quested et al. 2013).

From a methodological standpoint, this chapter, informed by qualitative and secondary data, does not contribute concretely to debates over which variables increase or decrease food waste. Instead, it explains food waste by looking beyond individuals and the household. Research that quantifies food waste and attempts to connect such data to specific purchasing patterns and lifestyles would therefore be valuable, as demonstrated in Chapters 10 and 12 in this collection, on food waste in Malaysia and the Philippines respectively. Such research would also contribute towards biophysical studies of food waste and improve the linkages between material and energy accounts of urban systems and the social, cultural and economic processes they embody.

In terms of policy, the connections among the food retail system, everyday life and household food waste that I have described suggest that more attention needs to be paid to these issues in general, but particularly in the rapidly developing cities of the Global South. There, the ongoing processes of urbanization and retail modernization suggest that there may still be opportunities for urban planners and other policymakers to shape the development of urban SOPs in ways that engender more sustainable forms of everyday life. This chapter reinforces the case for continued consumer education and practical training about best practices in food management, but also highlights the significance of SOPs. Further research is required to support such policies, but they might include regulating packaging and marketing tactics, or ensuring neighborhood walkability and convenient access to food to obviate stockpiling. Pushing the envelope a bit further, planners could also consider their role in developing sharing economies for food, either through support for community kitchens for different demographic groups, or infrastructure that allows the exchange of surplus food without the social stigma traditionally associated with food banks. Further research on the role of diversity in food retailing and its connections with the notions of mobility and time constraint could also provide insights into how people's lifestyles and everyday practices depend on and interact with the broader food system and other SOPs. Ultimately, the variety and complexity of the circumstances under which food becomes waste suggests that food waste requires wide-ranging and multi-scalar policy solutions that are highly context specific.

Notes

1 Interviewees were recruited via personal networks and online advertisements. Of the thirteen interviewees, only one was male. They ranged in age from twenty to sixty years old, and included housewives, full-time workers and university students.
2 This translates to approximately 75 kg of household food waste per capita. Although comparisons are tricky due to definitional and methodological differences, UK and US households threw away about 110 kg and 130 kg per capita in 2012 and 2010, respectively (WRAP 2013; Buzby et al. 2014).
3 Director, Ministry of Environment, Republic of Korea.
4 Also known as superstores, supercenters or large discount retailers, hypermarkets combine the sale of groceries with a wide variety of other goods, including electronics, appliances, clothing and sports goods in stores with large floor areas. Korean law defines hypermarkets as stores having floor areas larger than 3300 square meters.
5 Tesco sold Homeplus to a private investor consortium in September 2015, but this has been attributed more to Tesco's financial difficulties than the poor performance of Homeplus.
6 Defined according to Korean law as having floor area between 330 and 3300 square meters.
7 Defined according to Korean law as having floor area less than 330 square meters.
8 I have used pseudonyms for all my interviewees to preserve anonymity.
9 All of these interviewees were women; it is possible that the primary market for convenience and small-portion vegetables is men living alone.

References

Adhikari, B., S. Barrington, and J. Martinez. 2009. 'Urban food waste generation: Challenges and opportunities.' *International Journal of Environment and Waste Management* 3(1): 4–21. doi:10.1504/IJEWM.2009.024696.

Bak, S. 2014. 'Food, gender, and family network in modern Korean society.' *Asian Journal of Humanities and Social Studies* 2(1). http://ajouronline.com/index.php?journal=AJHSS&page=article&op=view&path%5B%5D=842

Bell, D., D. Corsten and G. Knox. 2011. 'From point of purchase to path to purchase: How preshopping factors drive unplanned buying.' *Journal of Marketing* 75(1): 31–45. doi:10.1509/jmkg.75.1.31.

Bluedorn, A. 2002. *The Human Organization of Time: Temporal Realities and Experience*, 1st edn. Stanford, CA: Stanford Business Books.

Brislin, R. and E. Kim. 2003. 'Cultural diversity in people's understanding and uses of time.' *Applied Psychology* 52(3): 363–382. doi:10.1111/1464-0597.00140.

Buzby, J., H. Wells and J. Hyman. 2014. 'The estimated amount, value, and calories of postharvest food losses at the retail and consumer levels in the United States'. Economic Information Bulletin 121. United States Department of Agriculture. www.ers.usda.gov/media/1282296/eib121.pdf (accessed 8 June 2016).

Chang, Y. and S. Lee. 2012. 'Does maternal employment affect parental time allocated to children's food consumption and physical activity? Evidence from the Korean Time Use Survey.' *International Journal of Human Ecology* 53: 67.

Cho, J., H. Chun and Y. Lee. 2013. 'Entry of large discount stores and the evolution of employment in the Korean retail sector.' www.frbatlanta.org/documents/news/conferences/13caed/F_2_Chun.pdf (accessed 8 June 2016).

Choe, S. 2012. 'South Koreans push back against giant hypermarkets.' *The New York Times*, 3 October, sec. Business Day/Global Business. www.nytimes.com/2012/10/04/business/global/battling-the-retail-goliaths-in-south-korea.html (accessed 8 June 2016).

Coe, N. M. and Y.-S. Lee. 2006. 'The strategic localization of transnational retailers: The case of Samsung-Tesco in South Korea.' *Economic Geography* 82(1): 61–88. doi:10.1111/j.1944-8287.2006.tb00288.x.

Coe, N. and Y.-S. Lee. 2013. '"We've learnt how to be local": The deepening territorial embeddedness of Samsung–Tesco in South Korea.' *Journal of Economic Geography* 13(2): 327–356. doi:10.1093/jeg/lbs057.

Cohen, D. and S. Babey. 2012. 'Contextual influences on eating behaviours: Heuristic processing and dietary choices.' *Obesity Reviews* 13(9): 766–779. doi:10.1111/j.1467-789X.2012.01001.x.

Distribution Development Industry Act. 2012. http://elaw.klri.re.kr/eng_mobile/viewer.do?hseq=16323&type=part&key=28 (accessed 8 June 2016).

Euromonitor. 2013. 'Grocery retailers in South Korea.' Passport. Euromonitor International.

Euromonitor. 2014. 'Grocery retailers in South Korea.' Passport. Euromonitor International.

Evans, D. 2012. 'Beyond the throwaway society: Ordinary domestic practice and a sociological approach to household food waste.' *Sociology* 46(1): 41–56. doi:10.1177/0038038511416150.

Evans, D. 2014. *Food Waste: Home Consumption, Material Culture and Everyday Life*. London: Bloomsbury.

Fine, B. and E. Leopold. 2002. *The World of Consumption: The Material and Cultural Revisited*, 2nd edn. London: Routledge.

Gallup Korea. 2015. 'Survey on changes in eating lifestyles, cooking shows, and nutrition from 1994–2015.' Gallup Korea Daily Opinion. http://panel.gallup.co.kr/Contents/GallupReport/한국갤럽GallupReport(20150820)_식생활변화와쿡방,보양식.pdf (accessed 8 June 2016).

Gandolfi, F. and P. Strach. 2009. 'Retail internationalization: Gaining insights from the Wal-Mart experience in South Korea.' *Review of International Comparative Management* 10(1): 187–199.

Godfray, H., J. Beddington, I. Crute, L. Haddad, D. Lawrence, J. Muir, J. Pretty, S. Robinson, S. Thomas, and C. Toulmin. 2010. 'Food security: The challenge of feeding 9 billion people.' *Science* 327(5967): 812–818. doi:10.1126/science.1185383.

Graham-Rowe, E., D. Jessop and P. Sparks. 2014. 'Identifying motivations and barriers to minimising household food waste.' *Resources, Conservation and Recycling* 84 (March): 15–23. doi:10.1016/j.resconrec.2013.12.005.

Gustavsson, J., C. Cederberg, U. Sonesson, R. Van Otterdijk and A. Meybeck. 2011. 'Global food losses and food waste: Extent, causes and prevention'. Rome: United Nations Food and Agriculture Organization. www.fao.org/ag/ags/ags-division/publications/publication/en/c/74045/ (accessed 8 June 2016).

Halkier, B. 2010. *Consumption Challenged: Food in Medialised Everyday Lives.* Surrey, UK: Ashgate.

Hamermesh, D. and J. Lee. 2007. 'Stressed out on four continents: Time crunch or yuppie kvetch?' *Review of Economics and Statistics* 89(2): 374–383. doi:10.1162/rest.89.2.374.

Jabs, J. and C. Devine. 2006. 'Time scarcity and food choices: An overview.' *Appetite* 47(2): 196–204. doi:10.1016/j.appet.2006.02.014.

Jin, B. and J.-O. Kim. 2003. 'A typology of Korean Discount shoppers: Shopping motives, store attributes, and outcomes.' *International Journal of Service Industry Management* 14(4): 396–419. doi:10.1108/09564230310489240.

Jun, M.-J., J. Kim, J. Kwon, and J.-E. Jeong. 2013. 'The effects of high-density suburban development on commuter mode choices in Seoul, Korea.' *Cities* 31 (April): 230–238. doi:10.1016/j.cities.2012.06.016.

Kim, B.-E. 2013. 'Single-person households on the rise.' *The Korea Times*, 13 March. www.koreatimes.co.kr/www/news/biz/2013/10/330_131393.html (accessed 8 June 2016).

Kim, K.-Y, and Y. Kim. 2012. '2011 modularization of Korea's development experience: Volume-based waste fee system in Korea'. Seoul: Korea Environmental Institute.

Kim, R. 2008. 'Wal-Mart Korea: Challenges of entering a foreign market.' *Journal of Asia- Pacific Business* 9(4): 344–357. doi:10.1080/10599230802453604.

Kim, S.-H. 2013. 'Korean retail giants run new types of stores to get around regulations.' *The Korea Herald*, 12 June. www.asianewsnet.net/Korean-retail-giants-run-new-types-of-stores-to-ge-47846.html (accessed 8 June 2016).

Kim, W. and A. Hallsworth. 2015. 'Tesco in Korea: Regulation and retail change.' *Tijdschrift Voor Economische En Sociale Geografie*, March. doi:10.1111/tesg.12145.

Kim, Y.-K. and Son, Y.-J. 2014. 'Prospects of Family Changes and Policy Implications'. Sejong-si: Korea Institute for Health and Social Affairs. https://www.kihasa.re.kr/common/filedown.do?seq=18610 (accessed 20 July 2016).

Koivupuro, H.-K., H. Hartikainen, K. Silvennoinen, J.-M. Katajajuuri, N. Heikintalo, A. Reinikainen and L. Jalkanen. 2012. 'Influence of socio-demographical, behavioural and attitudinal factors on the amount of avoidable food waste generated in Finnish households.' *International Journal of Consumer Studies* 36(2): 183–191. doi:10.1111/j.1470-6431.2011.01080.x.

Lee, H.-S., K. Duffey and B. Popkin. 2012. 'South Korea's entry to the global food economy: Shifts in consumption of food between 1998 and 2009.' *Asia Pacific Journal of Clinical Nutrition* 21(4): 618–629.

Lee, M.-J., B. Popkin and S. Kim. 2002. 'The unique aspects of the nutrition transition in South Korea: The retention of healthful elements in their traditional diet.' *Public Health Nutrition* 5(1a): 197–203. doi:10.1079/PHN2001294.

Lee, S., D. McCann, and J. Messenger. 2007. *Working Time around the World: Trends in Working Hours, Laws and Policies in a Global Comparative Perspective*. Routledge Studies in the Modern World Economy. Oxon, UK: Routledge.

Lee, Y.-O. 2005. 'Perceptions of time in Korean and English.' *Human Communication* 12(1): 119–138.

Ministry of Environment. 2014. 'Waste prevention policy.' http://eng.me.go.kr/eng/web/index.do?menuId=139&findDepth=1 (accessed 8 June 2016).

Moon, S. 2010. 'A study on determining the extent of damage and decreasing rate of supermarket sales by new SSM store : Using the Huff model.' Master's thesis, Seoul, Republic of Korea: Konkuk University. www.riss.kr/search/detail/ssoSkipDetailView.do?p_mat_type=be54d9b8bc7cdb09&control_no=40895ba4419eced3ffe0bdc3ef48d419 (accessed 8 June 2016).

Oum, Y. 2005. 'Authenticity and representation: Cuisines and identities in Korean-American diaspora.' *Postcolonial Studies* 8(1): 109–125. doi:10.1080/13688790500134380.

Quested, T., E. Marsh, D. Stunell and A. Parry. 2013. 'Spaghetti soup: The complex world of food waste behaviours.' *Resources, Conservation and Recycling* 79: 43–51.

Reardon, T., C. Timmer, C. Barrett and J. Berdegué. 2003. 'The rise of supermarkets in Africa, Asia, and Latin America.' *American Journal of Agricultural Economics* 85(5): 1140–1146. doi:10.1111/j.0092-5853.2003.00520.x.

Reckwitz, A. 2002. 'Toward a theory of social practices: A development in culturalist theorizing.' *European Journal of Social Theory* 5(2): 243–263. doi:10.1177/13684310222225432.

Røpke, I. 2009. 'Theories of practice—New inspiration for ecological economic studies on consumption.' *Ecological Economics* 68(10): 2490–2497.

Shove, E. 2003. 'Converging conventions of comfort, cleanliness and convenience.' *Journal of Consumer Policy* 26(4): 395–418. doi:10.1023/A:1026362829781.

Shove, E. 2009. 'Everyday practice and the production and consumption of time.' In *Time, Consumption and Everyday Life Practice, Materiality and Culture*, edited by E. Shove, F. Trentmann and R. Wilk, pp. 17–33. Oxford and New York: Berg.

Shove, E. 2010. 'Beyond the ABC: Climate change policy and theories of social change.' *Environment and Planning A* 42(6): 1273.

Shove, E. and D. Southerton. 2000. 'Defrosting the freezer: From novelty to convenience: A narrative of normalization.' *Journal of Material Culture* 5(3): 301–319. doi:10.1177/135918350000500303.

Shove, E. and G. Walker. 2010. 'Governing transitions in the sustainability of everyday life.' *Research Policy* 39(4): 471–476. doi:10.1016/j.respol.2010.01.019.

Song, J.-A. 2012. 'Costco vs Seoul: When Sunday comes…' *Beyondbrics (Financial Times)*, 5 October. http://blogs.ft.com/beyond-brics/2012/10/05/costco-vs-seoul-when-sunday- comes/ (accessed 8 June 2016).

Southerton, D. 2006. 'Analysing the Temporal organization of daily life: Social constraints, practices and their allocation.' *Sociology* 40(3): 435–454. doi:10.1177/0038038506063668.

Southerton, D. 2012. 'Habits, routines and temporalities of consumption: From individual behaviours to the reproduction of everyday practices.' *Time & Society*, December. doi:10.1177/0961463X12464228.

Southerton, D., H. Chappells and B. Van Vliet (eds). 2005. *Sustainable Consumption: The Implications of Changing Infrastructures of Provision*. Cheltenham: Edward Elgar.

Strasser, S. 2000. *Waste and Want: A Social History of Trash*, 1st edn. New York: Holt Paperbacks.

Suh, Y.-G. and E. Howard. 2009. 'Restructuring retailing in Korea: The case of Samsung-Tesco.' *Asia Pacific Business Review* 15(1): 29–40. doi:10.1080/13602380 802399312.

'Tesco Opens Worlds First Virtual Store.' 2011. *Tesco PLC*. 25 August. www.tescoplc.com/index.asp?pageid=17&newsid=345 (accessed 8 June 2016).

Venkat, K. 2012. 'The climate change and economic impacts of food waste in the United States.' *International Journal on Food System Dynamics* 2(4): 431–446.

Warde, A. 2005. 'Consumption and theories of practice.' *Journal of Consumer Culture* 5(2): 131–153. doi:10.1177/1469540505053090.

Warde, A., E. Shove, and D. Southerton. 1998. 'Convenience, schedules and sustainability.' In *European Science Foundation Workshop on Consumption, Everyday Life and Sustainability*. Lancaster, UK. www.lancaster.ac.uk/fass/projects/esf/convenience.htm (accessed 8 June 2016).

Watson, M. 2012. 'How theories of practice can inform transition to a decarbonised transport system.' *Journal of Transport Geography* 24 (September): 488–496. doi:10.1016/j.jtrangeo.2012.04.002.

Williams, H., F. Wikström, T. Otterbring, M. Löfgren and A. Gustafsson. 2012. 'Reasons for household food waste with special attention to packaging.' *Journal of Cleaner Production* 24 (March): 141–148. doi:10.1016/j.jclepro.2011.11.044.

Wilson, D., L. Rodic, A. Scheinberg, C. Velis, and G. Alabaster. 2012. 'Comparative analysis of solid waste management in 20 cities.' *Waste Management & Research* 30(3): 237–254. doi:10.1177/0734242X12437569.

WRAP. 2013. 'Household food and drink waste in the United Kingdom 2012.' www.wrap.org.uk/sites/files/wrap/hhfdw-2012-main.pdf.pdf (accessed 8 June 2016).

12 Towards sustainable consumption of rice in a private school in Metro Manila

Abigail Marie T. Favis and Rafael Deo F. Estanislao

With over four million hectares of arable land devoted to rice production and with an annual total harvest of 18,032,400 (palay[1]) metric tons, the Philippines is known to be the eighth largest producer of rice in the world (Global Rice Science Partnership 2013: 33). Though rice has become a staple food in everyday Philippine diets today, it was not always widely accessible. In pre-colonial times, rice was considered to be a 'prestigious and highly valued food' and was often used as tributes or gifts for royalty (Aguilar 2005: 2). Over time, as social and economic systems evolved, rice became more available, accessible and central to Filipino food culture. During the 1980s and 1990s, the annual per capita consumption of rice was estimated to be between 90 to 92 kg. Around two decades later, consumption has increased to 111 kg per person (Lantican et al. 2013: 11). While the Philippines is one of the top ten rice producers, it is also a rice importer with around 20 percent of its local consumption supplied by imported rice (Wailes and Chavez 2012: 9). Rice imports have increased in the past ten years, brought about by a growing population and rising income levels (Department of Agriculture 2012: 2).

Rice has always been the main focus of agricultural policy – with much of the attention given to promoting rice self-sufficiency and regulating pricing. One of the national government's initiatives to ensure a rice self-sufficient future for the Philippines is the Food Staples Sufficiency Program (FSSP) implemented by the Department of Agriculture. A FSSP key strategy is to reduce rice losses in both the post-harvest and post-consumer stages (Department of Agriculture 2012: 33, 40). In support of this, the government declared 2013 as the National Year of Rice (Proclamation No. 494 2012). While the theme of the celebration was '*Sapat na Bigas, Kaya ng 'Pinas* (the Philippines can be rice sufficient)', the issue of rice wastage was also given much attention. Various public and private agencies and non-governmental organizations have already focused on reducing rice losses throughout the value chain. Research conducted by the Philippine Rice Research Institute (PhilRice) and the Philippine Center for Post-Harvest Development and Mechanization (PhilMech) show that around 16.47 percent of domestic harvest is lost mostly during drying and milling with minimal losses incurred in piling/handling, threshing and storage (House of Representatives 2015: 1). Such losses have been attributed to outdated or improper methods (for example, road-side or uneven drying) or poorly operated machinery. Efforts are already in place to address these losses through the establishment of modern, integrated rice processing complexes throughout the country (Philippine Center for Post-Harvest Development and Mechanization 2012: 47). These installations minimize post-harvest losses through the use of modern techniques, leading to greater efficiency in operations.

The Food and Nutrition Research Institute (FNRI) has been including post-consumer household rice waste in its National Nutrition Surveys. Their results show that per capita rice waste (white, milled rice) among Philippine households is equivalent to 9.0 g or roughly two tablespoons per household daily (FNRI 2012: 2). This figure does not include rice waste generated in institutions and organizations, for which the FNRI has no data. As such, little detail is known about the amounts of rice wasted in the post-consumer phase. This is significant information that may inform future policies related to food waste minimization. In addition, while it has been reported that most food losses in developing countries occur during the post-harvest stages of the supply chain (Lipinski et al. 2013: 8; FAO 2013: 14), new evidence shows that cities in India and China are exhibiting significant losses at the consumption end (Teng and Trethewie 2012: 2), which may point to similar changes in cities in developing economies. Hence, it is worthwhile to investigate post-consumer food wastage in cities and urban centers such as Metro Manila.

Households have often been the focus of food consumption and food waste studies, yet less attention is given to the meso-level, including restaurants, the workplace, hospitals and schools (see Chapter 10 in this volume for a discussion around restaurant waste in Kuala Lumpur). Exploring food waste generation at the level of institutions is highly relevant, as this allows for an understanding of practices regarding food on a scale much larger than the household. For instance the drivers for food waste generation in households may be different from the drivers of waste generation in hospital canteens or mall food courts. Thus, approaches to food waste avoidance or minimization must take into account different contexts and scales in order to achieve successful results. Differences in the practices people have regarding food in a household as compared to their practices with food in large, perhaps communal settings may mean that different approaches may be needed to avoid or minimize waste.

Educational institutions are particularly interesting to investigate as these play a crucial role in promoting sustainability. As institutions for education and formation, schools are in a prime position to provide knowledge, introduce or reinforce norms, and demonstrate practices regarding sustainable food consumption. Literature on food waste generation in schools covers a wide array of related topics from quantification/characterization (Williamson et al. 2003; Cohen et al. 2013; Hoy et al. 2013; Hanks et al. 2014; Martins et al. 2014) to the assessments of intervention methods (Marlette et al. 2005; Byker et al. 2014; Cullen et al. 2015; Goldberg et al. 2015). Methods for measuring food waste in school canteens include direct weighing, visual estimation, photograph analysis and surveys. The method used in this study is direct weighing, acknowledged to be the most accurate method (Martins et al. 2014). While food waste is defined in varied ways (Parfitt et al. 2010; Garrone et al. 2014), this study uses the definition of the FNRI, which defines it as 'cooked or raw food items that a person or family failed to consume due to cooking, wastage from cooking and plate, and fed to animals and pets' (Balitaon, n.d.). This study measured plate food waste in a private educational institution in order to assess rice wastage in school lunch servings.

To understand how and in what way wastage occurs, we begin by exploring social practice theory in relation to rice waste generation in the section that follows. We then introduce our methodology and data collection methods, including uncertainties. In the third section, we present the results of our case study, followed by an analysis of our data. We conclude with a discussion around the significance of combining a social and biophysical understanding of how rice waste is generated and can be reduced.

Social practice theory in the context of rice waste generation

Early models of understanding human behavior, such as Azjen's Theory of Planned Behavior (Azjen 1991), considered an individual's behavior as being shaped by attitudes, subjective norms and perceived behavioral control. Over the years, many other factors and variables – such as belief salience, past habits, moral norms and the like – have been added to the model in order to more fully understand how an individual's behavioral choices are formed (Hargreaves 2011: 80). This has resulted in increasing model complexity, diminishing its ease of application. In addition, certain models remain highly individualistic and fail to take into consideration other factors outside the self that influence behavior.

In contrast, social practice theory views human activities as influenced by social elements, both individual and collective, and material aspects. The most oft-quoted definition of social practice theory by Reckwitz is as follows:

> a routinized type of behavior which consists of several elements, interconnected to one other: forms of bodily activities, forms of mental activities, 'things' and their use, a background knowledge in the form of understanding, know-how, states of emotion and motivational knowledge.
> (Reckwitz 2002: 249)

Practices such as eating in a canteen, dining out, shopping for food and getting food to-go, are a 'routinized way in which bodies are moved, objects are handled, subjects are treated, things are described' (Reckwitz 2002: 249). This, in turn, relates to habits which are 'recurrently and consistently reproduced by suitably committed practitioners' (Shove 2012: 103). Habits are deeply ingrained in three aspects of practice – the body, the material world, and the social world – and are difficult to modify (Sahakian and Wilhite 2013). The body refers to routinized bodily activities, including mental and emotional activities (Reckwitz 2002: 251). It includes 'cognitive processes and physical dispositions, acquired by the body through social experiences, inscribed in space and over time' (Sahakian and Wilhite 2013: 28). The material world refers to 'social practices [that] consist of routinized relations between several agents ... and objects' (Reckwitz 2002: 253). As discussed in Sahakian and Wilhite (2013), our dispositions and therefore our practices are affected by the material dimension in which we interact; infrastructure and technology act upon our own actions. Carrying out a practice requires us to act on objects in a certain way (Spaargaren 2011: 817). The social context of consumption practices requires that attention be given to social rules and values related to consumption. These social rules and values can be contested and possibly changed when they are brought into open debate and discussion (Sahakian and Wilhite 2013: 30).

If change towards more sustainable behaviors is desired, how these three aspects of practice enforce or sustain habits and routines must be understood. This opens up opportunities for social learning – which is a combination of both new information and engagement in new practice, a marriage of individual experience and collective social participation. The following case study presents the problem of rice waste generation in school cafeterias and the initiatives implemented to address it. Social practice theory as applied in this context is used to further understand rice waste generation among diners and the interventions that may influence changes towards less wasteful practice. While personal preferences for taste, food quality and other bodily aspects may be important dimensions in understanding why students do not finish their food, investigating the material and social dimensions of dining practices is also significant, as we will demonstrate.

Methodology, data collection and uncertainties

Research was conducted in a private school in Metro Manila in August 2013, February 2014 and September 2014. The food waste of all canteen diners was investigated in the main school canteens, where the bulk of the school meals are served, for the primary (grades one through eight) and tertiary (university) education levels, which corresponds to students aged eight to sixteen years for the primary level, and sixteen to twenty-two years for the tertiary level. The school under study is a private institution, with an all-boys basic education (primary and secondary school) and a co-educational tertiary unit. A common methodology was followed for all canteens. In order to maximize the data collected during the audits, other food categories such as meat/poultry/fish and vegetables/fruits (and pasta/pastry for the tertiary level) were also measured, in addition to rice. The waste audits were not publicized and were conducted as discreetly as possible, so as to not influence or interfere with the existing dining practice. This was done to minimize bias or uncertainty in gathering baseline food waste data.

Prior to the conduct of the food waste audits, consultation meetings with administrators, canteen personnel, food safety officers, and student leaders were conducted to ensure the support and cooperation of all stakeholders. Such consultations are part of the school's consultative and inclusive approach in implementing new programs or projects, as opposed to a top-down directive. Ensuring that stakeholders understand the goals and objectives of the activity does not only ensure that implementation of the initiatives is smoother, it also provides an opportunity for them to express their ideas, suggestions and concerns, which helped improve the methodology. All canteen personnel and student volunteers involved were especially briefed on the goals, objectives and methodology of the waste audit.

Five regular school days (i.e. days with no special events such as food sales and extracurricular activities) were randomly chosen as the duration of the food waste audits in both units. The audit in the primary school was conducted during the regular lunch break schedule. The audit in the tertiary level was conducted on three main canteens during peak dining hours, as there is no set schedule for lunch. During the sampling period, pre-weighed and labeled segregation bins or pans were provided in strategic locations within the canteens, as close to the existing collection and segregation stations as possible. Diners were requested to separate their plate waste into rice, meat/poultry/fish, vegetables/fruit, and for the tertiary unit, pasta/bread. Collectors and monitors were tasked to ensure that the segregation process was as complete as possible. Collectors and monitors also observed the behavior of canteen diners and staff at all stages of the data collection, and recorded their observations. Student volunteer observations also contributed to our research analysis. Informal interviews were conducted with students and staff.

At the end of the daily audit period, the sorted food wastes were weighed. The number of plates, representing the number of diners, was also documented. At the end of the sampling period, per capita wastage and average daily waste generation were calculated. The food waste audits were conducted by volunteers from among administrators, faculty members, students and canteen staff, who were willing and cooperative, after several consultations that involved engaging them in planning and training. This level of engagement ensured sufficient administrative support for the endeavor.

Given the limited duration of the waste audits, uncertainties in the data are to be expected. The standard deviation values of data obtained from both the primary school and tertiary unit indicate considerable day-to-day variability. In future studies, this may be addressed by lengthening the duration of the audit period and by sampling regular school days throughout the school year to account for seasonal variability, if any. Finally, it is also important to remember that this data only accounts for the rice waste from food plates served in the canteen and does not include the leftover rice from packed lunches from home. It has also not taken into consideration losses from the procurement, storage, preparation and cooking of rice, which may also be a significant contributor to total rice waste.

Case study results

The school administration recognizes that canteens are a significant source of food waste which is not only a solid waste issue, but also involves resource consumption and related environmental impacts upstream, such as the loss of land, water and energy (FAO 2011). Through life-cycle analysis – a technique to assess the environmental aspects and impact of a product or process – it is recognized that waste food translates to wasted water and other material and energy inputs, during food cultivation, processing and transport. In recent years, the school has begun to implement sustainable campus initiatives, which includes a Sustainable Food Program. This program aims to render the school's food planning, purchasing, consumption and waste generation activities as socially, environmentally and financially responsible as possible. In order to manage this, baseline information was needed to serve as a point of reference and comparison. Thus, school canteens in the primary (grades one through eight) and tertiary (university) levels were studied.

In the primary school canteen, the students who wish to avail of canteen lunches line up at the counter to receive their lunch food plates – which consist of rice, meat, vegetables and fruit. Since the early 2000s, the primary school canteen served unlimited portions of rice, with the intent of providing more nutrition and energy for the growing children. In addition to the rice gained at the counter, rice bowls were also placed on each dining table; students could freely help themselves to more servings. Such passive consumption was not given serious thought until the declaration of the National Year of Rice in 2013, a topic we will come back to later in this chapter. In the tertiary unit,

Figure 12.1 The food waste collection bin and the plastic trash bins.
Source: Marlyne Sahakian.

unlike the primary school, there are more choices available for university students to avail of, including pasta and bread. The canteens do not provide unlimited rice but some concessionaires provide an option for half rice servings. Similar to the primary school, the question of rice portions and servings was not given much attention until the declaration of the National Year of Rice in 2013.

In all canteens studied, it is common practice for students to properly dispose of their leftover food in appointed waste bins. From there, the kitchen staff take charge of disposing of the collected food waste either through diversion to existing on-campus composting facilities or through an accredited waste collector. All school canteens are managed by accredited contractors, and not by the school itself.

Primary school results

An average of 611 diners ate at the primary school canteen daily during the audit period. While the total population is around 4200, most students have packed lunches which they bring home at the end of the day. Data collected show that most of the plate waste is rice (49 percent), followed by meat/poultry/fish (41 percent), then to a lesser extent vegetables/fruit (10 percent). The total rice waste measured during the sampling period amounted to 50.4 kg, at an average of 10.1 kg a day. Empirical measurement of rice serving size in the canteen indicates that one serving is approximately 175 g. Thus, the total wasted rice for the school week is equivalent to around 288 servings. Per capita rice waste was found to be 16.5 g per day for the lunch period alone. This is approximately 83 percent more than the FNRI data of 9.0 g per capita per day, which involves waste throughout the day and not solely for one meal. These results are significant, not only because rice is a contentious and political commodity, but also because the volume of rice wasted also points to issues related to existing consumption practices. Are students indifferent to wasting rice, or are other factors at play?

Based on our observations and interviews with the school's Food Safety and Quality Assurance Supervisor, rice waste generation seems to be influenced by two main factors. First, when students are served the one cup of rice included in their meal, they may also help themselves from a bowl of rice on their dining tables, which may be encouraging them to over-serve themselves rice. Second, the serving cup used to measure the rice is the same for children and adults. This means that the portioning of rice given to the younger students might be excessive in the first place.

Per capita and combined meat, poultry and fish waste was found to be 13.7 g per capita. This value is much larger (77 percent) than the FNRI data for per capita meat and poultry waste, which is 3.0 g. Vegetable and fruit waste was found to be 3.2 g. This value is close to the FNRI data for per capita vegetable/fruit waste, which is 2.0 g. It must be noted that the food plate waste figures for meat, poultry and fish on the one hand, and vegetables and fruits on the other, also account for the unavoidable waste portions such as bones, peel and seeds, among others. Such inedible parts are referred to as unavoidable food waste (Papargyropoulou et al. 2014; Papargyropoulou, Chapter 10, this volume); this distinction is central to identifying interventions that may decrease food waste generation.

University level results

An average of 1573 diners ate at the tertiary unit canteens daily during the audit period. These canteens involve a vast number of dining opportunities, or food concessionaires offering everything from Filipino fare to hotdogs and pastries. This number does not take into account the portion of the population that bought food from the canteen to eat and dispose of elsewhere. This also does not include the portion of the population that brought packed food from home. Data collected show that most of the plate waste is rice (41 percent), followed by meat/poultry/fish (28 percent), vegetables/fruit (24 percent), and pasta/pastry (7 percent). Total rice wasted during the audit period amounted to 40.0 kg. This value is equivalent to around 229 servings of rice wasted. This is less than the amount of food wasted in the primary school, possibly because there is no unlimited rice option at the tertiary level and university canteens also offer other starch options, such as pasta and pastries. Per capita rice wastage is 5.1 g per day for lunch alone, which is less than the national average of 9.0 g – which includes rice waste generated during the entire day, not solely during lunchtime.

Based on informal interviews and observations, perceptions of rice quality had a role to play in wastage. Students explained that cooked rice quality varies depending on which concessionaire it is bought from – the rice can be perceived as too dry, too mushy, have a fragrant or a less palatable scent. These differences

Figure 12.2 The university canteen.
Source: Marlyne Sahakian.

influence their decision to finish or not finish the rice serving. Dining is also often accompanied by other activities. Individual diners can be seen studying or working on computers right after meals. For the larger groups, dining together is also a social affair. It is common to see and hear loud conversations, card games and discussions around group dining. On occasion, smaller groups also get together in meetings to discuss academic or extra-curricular activities. University-level diners have flexible schedules and therefore spend more time lingering at their tables and may therefore have more opportunities to finish their meals.

In addition, more university students buy from food stalls that provide set meals (such as a meat, vegetable and rice dish, or a hamburger), as opposed to the main concessionaire that sells meal components separately. The vegetable components of the set meals are, in almost all cases, stir fried beansprouts. A diner may choose to not have this side dish, but must still pay the set price. Or, the diner may also choose to substitute available options such as eggs or noodles. The rice portion may also be adjusted, should the diner prefer less rice. Yet even if they halve the serving, they still must pay full price. At the main concessionaire and for university students, diners are provided the option to avail of half servings for rice, at a reduced price.

Per capita meat/poultry/fish waste generation at the university level is 3.5 g. This is close to the national average of 3.0 g. Per capita vegetable/fruit waste generation is 3.0 g, which is also close to the national average. Per capita pasta/pastry waste generation is 0.82 g.

Analysis: existing and new practices related to rice consumption and waste

Students' rice-eating habits consist of routinized practices involving several elements, including type, quality and quantity of rice available, as well as established norms in canteen settings. Actual serving sizes can be pre-determined from a nutritional standpoint, but the age of the consumer must also be taken into account – in terms of the size of measuring and serving implements, such as rice scoops (Barba et al. 2008). In addition, the presence of rice bowls on the dining tables from which students can get additional rice servings freely may have created a setting where it is seen as normal to get as much rice as possible: if the rice is freely available on the table, surely it should be consumed. Such passive consumption of rice may be prompting students to serve themselves more rice than they can actually finish. This practice was accepted as the norm and was not monitored by the adult supervisors who accompany the primary school students during lunch breaks. Students were also not given any reminders or prompts for less wasteful food consumption, thus the over-serving of rice and disposing of rice leftovers were not top of the mind.

Results of the plate waste audit show that per capita, less rice waste is generated in the tertiary level – which may be explained by more options for other cereals and starches, such as pasta and bread, which are not commonly available in the primary level. While unlimited rice options were not available at the tertiary-level canteens, there was nevertheless rice wastage due to routinized bodily activities, including mental and emotional activities learned over time, and the material setting of food consumption. Informal interviews with the students indicate three possible reasons why rice was wasted: large portion size, financial factors and the quality of the cooked rice. Portion size plays a common role when asked why they do not finish their food. For example, some meals from concessionaires are composed almost entirely of rice, as their menu consists of a variety of fried rice and is not adapted to the appetite of the customer. Financial factors, such as having to pay full price even for a half-serving of rice, discourage students from choosing a smaller portion – as this practice is not aligned with the norm of seeking 'value for money'. In addition, as rice preparation is not standardized and different rice varieties and cooking methods are used, the quality of the cooked rice itself may have influenced whether or not the students finish their servings.

Collectively and in all canteens, there also seems to be a lack of awareness around the true costs of rice waste. Information could play a role, including diners with appropriately designed information (with approachable language and relevant examples) on the socio-economic and environmental impacts of rice and food waste. For social learning to occur, providing information alone would not suffice: new opportunities to participate in less wasteful practices could also be designed, for example by reducing the amount of free rice available in the primary school canteen, and introducing half-rice servings priced accordingly at the university level.

Since this study was conducted, results of the study were presented to school administrators and efforts were made to reduce rice wastage, particularly in the primary school where free-serving rice bowls were removed from the tables. Instead, students must now approach the service counter should they wish for another serving of rice. Based on observations gathered since this new policy has been in place, few students requested extra cups of rice. One hindrance may be the additional steps required for an additional serving, involving having to walk back to the service counter, fall in line again, and wait to be given the additional serving. More students finished their meals at the table, with less rice waste generated. Thus, modifying the material aspect of consumption – or removing self-service rice bowls and offering second servings at the counter – contributed to the reduction of rice waste. In addition, teachers and lunch supervisors were asked to remind students to get only what food they felt they could finish, and to actually finish their meals. These efforts created a new norm around the belief that wasting rice is not desirable, reinforced by social learning opportunities including the school's Nutrition Week and ongoing discussions between students, teachers and moderators. While no official dialogue was held with the parents, several of them approached the office of the Food Safety and Quality Assurance Officer to thank the school for implementing new programs aimed at minimizing food waste, as they also work to encourage their children to finish their meals at home. The parents also asked for the school to continue to reinforce such behaviors in school, because this would affect household behavior A few meetings were held with parent representatives to discuss how students can also monitor and minimize their food waste at home. Thus, changes in the social context of rice consumption practices came about when social rules and values related to consumption were scrutinized and opened to discussion – with an interplay between school and home.

However, more needs to be done. There are many aspects yet to be addressed, such as the change in rice-serving portioning tools used for children and the pricing of half versus full servings of rice at the tertiary unit. While new practices are already being introduced, these should also be complemented by new information, but going beyond posters and pamphlets. Education around food waste could be integrated into the curriculum, as well as non-academic formation activities, focused particularly on an understanding of complex systems and the development of critical thinking skills among students, to respond to challenges of food waste and sustainable food consumption more generally. This topic could be discussed in a wide array of disciplines – from social studies to the natural sciences. Educational programs have already been shown to be effective in helping consumers make better decisions in terms of preventing food waste (Whitehair et al. 2013). Messaging and feedback campaigns, where people engage in two-way forms of communication, are also a tried and tested method in influencing consumers to make more sustainable choices. All these different factors could come together to reduce food waste.

Beyond the classroom, non-academic formation activities – such as extra-curricular projects, seminars, field trips and the like – are also important tools in bringing food waste issues to light and in challenging students to evaluate their attitudes towards food consumption. It has been shown that students acquire skills and knowledge informally through their life experiences both within and outside the campus (Barth et al. 2014). For example, heedless food waste generation might be due to the alienation of young Filipinos from the activities of food production and its related issues (Aguilar 2005). Thus, exposure to food production such as immersion trips to rice farms might help students gain a more visceral understanding of the environmental, social and economic aspects of growing rice, leading them to experience rice (and other food products) as more than just a consumer good.

Food plate waste is, however, just one part of the total waste generated by schools. Avoiding or minimizing food waste would need to occur upstream from diners' plates, through a concerted effort at the level of canteen operations. This is one of the biggest challenges in shifting towards sustainable food consumption at this private school, as the canteens are not managed by the school itself, but are contracted out to accredited parties. For these suppliers of canteen services, the economic cost of reducing waste could become a compelling argument. By studying food flows in canteen systems – from procurement, to storage, to food preparation to service – points for intervention may be identified, which could lead to increased resource-use efficiency and minimized waste, ultimately influencing profits positively.

Therefore, for the entire school to become more sustainable in food consumption and waste reduction, the entire organization – not just the students – has to change. Teachers could include issues related to rice and other food production in their classroom lessons and field trips. Closer coordination with parents could help ensure that efforts towards forming sustainable food behaviors and attitudes are streamlined between the school and the home. Advocacy campaigns, possibly spearheaded by environmental organizations and clubs within the school, could provide students and other campus users with information about the environmental, social and economic aspects of food wastage. Canteen management could improve the efficiency of their systems to maximize resources and minimize wastage, while administrators could strongly support and require that such activities mentioned above take place, including the monitoring and evaluation of results. All of these efforts together could include new opportunities for social learning and participating in new practices.

Conclusion: towards more sustainable food consumption

Schools are at a prime position to offer education of and values formation in resource conservation, waste minimization, stewardship, social responsibility and sustainability. They are also unique sites in which to implement change as they involve a centralized system of governance – the school administration – acting over a bounded territory that includes thousands of campus users, ranging from students to staff. Schools also offer an opportunity for learning new practices that can then lead to changes in other areas of consumption, including the home – in interplay between meso and micro scales of consumption.

In support of the National Year of Rice, and in line with its effort to build a more sustainable campus, a private school in Metro Manila determined the baseline plate waste generation in its canteens. Through this effort, we determined that per capita rice waste generation is 16.5 g per day at the primary school and 5.1 g per day at the tertiary unit. Understanding waste streams is the first step towards enhancing the sustainability of waste management systems (Smyth 2010) and gathering baseline information on plate waste generation is a crucial component in shifting to more sustainable food consumption in schools. The data gathered from the canteen food waste audits sheds light on *how much* food is wasted, in both the avoidable and unavoidable portions. Information about rice wastage is particularly significant, as it is the biggest component of the food waste stream in both the primary school and tertiary unit. However, it is also important to understand *why and in what way* waste is generated. By engaging with social practice theory, the interplay of the material world, routinized bodily practices, and the social context of rice (and food) waste generation was identified. These aspects were the focus of initial interventions to address a certain passivity around rice consumption and related waste among young students, yet more needs to be done. Actions already taken include information through an annual Nutrition Week celebration, which worked on the cognitive dimension and through social learning, as well as removal of free-serving bowls, which affected the material dimension of consumption. In the tertiary unit, measures are underway to ensure that all canteens are able to provide half-rice servings, with the corresponding reduction in price. The food safety and quality assurance office has also implemented more stringent food quality assessment measures to provide the canteens with feedback for improvement.

Future audits are already being planned to measure the impact of changes made so far at this private school. While the plate waste audit has brought to light much new information on dining behavior and practice, this data on waste is only from the post-consumer end. If a sustainable food program is to be implemented by the school, the whole food system must be studied. Thus, a more complete material flow analysis of food products in the school system is also currently underway, in order to have complete picture of resource flows within canteen systems and how these are transformed into waste. The scale of such actions at the level of a primary school and university campus show that efforts to reduce waste can be significant, and may contribute to a significant decrease food waste over a larger scale. Working at the meso-level of a campus has led to interplay between household and school consumption. A similar scale of food consumption is evident in other dining spaces, involving institutions such as hospital, government building or private workplace canteens, as well as places of concentrated activity, such as shopping malls. All of these spaces of consumption become relevant towards the Philippines' Food Staples Sufficiency Program, with an emphasis on understanding not only how much food is wasted, but also why and in what way practices could be shifted to more sustainable patterns. What we have shown through this study is that information and awareness is important, but changing social norms and the material dimension of consumption is also significant.

Note

1 Unhusked rice grains.

References

Aguilar, F., Jr. 2005. 'Rice in the Filipino diet and culture.' Philippine Institute for Development Studies (PIDS) Discussion Paper 2005-15. Makati City, Philippines: PIDS. http://dirp3.pids.gov.ph/ris/dps/pidsdps0515.pdf (accessed 11 June 2015).

Azjen, I. 1991. 'Theories of cognitive self-regulation: The theory of planned behavior.' *Organizational Behavior and Human Decision Processes* 50(2): 179–211. www.sciencedirect.com/science/article/pii/074959789190020T (accessed 20 September 2015).

Balitaon, M., A. Ducay, M. Constantino, I.-A Adeppa. n.d. 'Household plate waste among Filipino households and its socioeconomic characteristics.' Food and Nutrition Research Institute 40th Seminary Series. http://202.90.141.88/

Seminar%20Series/40th/Household%20Plate%20Waste%20Among%20Filipino%20Households%20.pdf (accessed 20 September 2015).

Barba, C. and M. Cabrera. 2008. 'Recommended energy and nutrient intakes for Filipinos 2002.' *Asia Pacific Journal of Clinical Nutrition* 17(S2): 399–404. http://apjcn.nhri.org.tw/server/apjcn/17/s2/399.pdf (accessed 28 September 2015).

Barth, M., M. Adomßent, D. Fischer, S. Richter and M. Rieckmann. 2014. 'Learning to change universities from within: A service-learning perspective on promoting sustainable consumption in higher education.' *Journal of Cleaner Production* 62: 72–81. www.sciencedirect.com/science/article/pii/S0959652613002072 (accessed 20 September 2015).

Byker, C., A. Farris, M. Marcenelle, G. Davis and E. Serrano. 2014. 'Food waste in a school nutrition program after implementation of new lunch program guidelines.' *Journal of Nutrition Education and Behavior* 46(5): 406–411. www.sciencedirect.com/science/article/pii/S1499404614001778 (accessed 20 September 2015).

Cohen, J., S. Richardson, S. Austin, C. Economos and E. Rimm. 2013. 'School lunch waste among middle school students: Nutrients consumed and costs.' *American Journal of Preventive Medicine* 44(2): 114–121. www.sciencedirect.com/science/article/pii/S074937971200760X (accessed 20 September 2015).

Cullen, K., T-A. Chen and J. Dave. 2015. 'Changes in foods selected and consumed after implementation of the new National School Lunch Program meal patterns in southeast Texas.' *Preventive Medicine Reports* 2: 440–443. www.sciencedirect.com/science/article/pii/S2211335515000674 (accessed 20 September 2015).

Department of Agriculture. 2012. *Food Staples Sufficiency Program 2011–2016*. Philippines: Department of Agriculture.

FAO. 2011. *Global Food Losses and Food Waste – Extent, Causes and Prevention*. Rome: Food and Agriculture Organization of the United Nations. www.fao.org/docrep/014/mb060e/mb060e.pdf (accessed 11 June 2015).

FAO. 2013. *Food Waste Footprint: Summary Report*. Rome: Food and Agriculture Organization of the United Nations. www.fao.org/docrep/018/i3347e/i3347e.pdf. (accessed 11 June 2015).

FNRI. 2012. *Digest Supplement*. Philippines: Department of Science and Technology. http://202.90.141.88/FNRI%20Digest/digest2012jan_mar.pdf (accessed 10 September 2015).

Garrone, P., M. Melacini and A. Perego. 2014. 'Opening the black-box of food waste reduction.' *Food Policy* 46(2014): 129–139. www.sciencedirect.com/science/article/pii/S0306919214000542 (accessed 20 September 2015).

Global Rice Science Partnership. 2013. *Rice Almanac*, 4th edn. Los Baños, Philippines: International Rice Research Institute. http://books.irri.org/9789712203008_content.pdf (accessed 11 June 2015).

Goldberg, J., S. Folta, M. Eliasziw, S. Koch-Weser, C. Economos, K. Hubbard, L. Tanskey, C. Wright and A. Must. 2015. 'Great taste, less waste: A cluster-randomized trial using a communications campaign to improve the quality of foods brought from home to school by elementary school children.' *Preventive Medicine* 74: 103–110. www.sciencedirect.com/science/article/pii/S0091743515000614 (accessed 20 September 2015).

Hanks, A., B. Wansink and D. Just. 2014. 'Reliability and accuracy of real-time visualization techniques for measuring school canteen tray waste: Validating the quarter-waste method.' *Journal of the Academy of Nutrition and Dietetics* 114(3):

470–474. www.sciencedirect.com/science/article/pii/S2212267213013373 (accessed 20 September 2015).

Hargreaves, T. 2011. 'Practicing behavior change: Applying social practice theory to pro- environmental behaviour change.' *Journal of Consumer Culture* 11(1): 79–99. http://joc.sagepub.com/content/11/1/79.short?rss=1&ssource=mfr (accessed 20 September 2015).

House of Representatives. 2015. *Facts in Figures No. 3: Rice Post-Harvest Losses and Consumption Wastage.* Congressional Policy and Budget Research Department, House of Representatives, Philippines. http://cpbrd.congress.gov.ph/images/PDF%20Attachments/Facts%20in%20Figures/FF2015-03%20Rice%20Postharvest.pdf (accessed 11 June 2015).

Hoy, K., A. Hanks, D. Just and B. Wansink. 2013. 'Visual inspection of canteen waste by quarter servings rivals the accuracy of weighing.' *Journal of Nutrition Education and Behavior* 45(4, Supplement): S47–S48. www.sciencedirect.com/science/article/pii/S1499404613003047 (accessed 20 September 2015).

Lantican, F., M. Sombilla and K. Quilloy. 2013. *Estimating the Demand Elasticities of Rice in the Philippines.* Monograph. Los Baños, Laguna: SEARCA. www.searca.org/knowledge-resources/1603-pre-download?Pid=232 (accessed 11 June 2015).

Lipinski, B., C. Hanson, J. Lomax, L. Kitinoja, R. Waite and T. Searchinger. 2013. *Reducing Food Loss and Waste – Installment 2 of 'Creating a Sustainable Food Future'.* Washington, DC: World Resource Institute. www.wri.org/sites/default/files/reducing_food_loss_and_waste.pdf (accessed 11 June 2015).

Marlette, M., S. Templeton and M. Panemangalore. 2005. 'Food type, food preparation, and competitive food purchases impact school lunch plate waste by sixth-grade students.' *Journal of the American Dietetic Association* 105(11): 1779–1782. www.sciencedirect.com/science/article/pii/S0002822305015464 (accessed 20 September 2015).

Martins, M., L. Cunha, S. Rodrigues and A. Rocha. 2014. 'Determination of plate waste in primary school lunches by weighing and visual estimation methods: A validation study.' *Waste Management* 34(8): 1362–1368. www.sciencedirect.com/science/article/pii/S0956053X14001299 (accessed 11 June 2015).

Papargyropoulou, E., R. Lozano, J. Steinberger, N. Wright and Z. bin Ujang. 2014. 'The food waste hierarchy as a framework for the management of food surplus and food waste.' *Journal of Cleaner Production* 76: 106–115. www.sciencedirect.com/science/article/pii/S0959652614003680 (accessed 20 September 2015).

Parfitt, J., M. Barthel and S. Macnaughton. 2010. 'Food waste within food supply chains: Quantification and potential for change to 2050.' *Philosophical Transactions of the Royal Society B* 365: 3065–3081. http://rstb.royalsocietypublishing.org/content/365/1554/3065 (accessed 20 September 2015).

Philippine Center for Post-Harvest Development and Mechanization. 2012. *Annual Report.* Philippines: Department of Agriculture. www.philmech.gov.ph/upload/PUBLICATIONS/2013/annual_report/annual%20report%202012.pdf (accessed 11 June 2015).

Proclamation No. 494, s. 2012. *Declaring the Year 2013 as the National Year of Rice and Directing the Department of Agriculture to Lead its Celebration.* www.gov.ph/2012/10/18/proclamation-no-494-s-2012/ (accessed 11 June 2015).

Reckwitz, A. 2002. 'Toward a theory of social practices: A development in culturalist theorizing.' *European Journal of Social Theory* 5(2): 243–263. http://est.sagepub.com/content/5/2/243.full.pdf+html (accessed 20 September 2015).

Sahakian, M. and H. Wilhite. 2013. 'Making practice theory practicable: Towards more sustainable forms of consumption.' *Journal of Consumer Culture* 14: 25. http://joc.sagepub.com/content/14/1/25 (accessed 20 September 2015).

Shove, E. 2012. 'Habits and their creatures.' In A. Warde and D. Southerton (eds), *The Habits of Consumption*. Helsinki: Helsinki Collegium for Advanced Studies.

Smyth, D., A. Fredeen and A. Booth. 2010. 'Reducing solid waste in higher education: The first step towards "greening" a university campus.' *Resources, Conservation and Recycling* 54(11): 1007–1016. www.sciencedirect.com/science/article/pii/S0921344910000492 (accessed 20 September 2015).

Spaargaren, G. 2011. 'Theories of practice: Agency, technology, and culture. Exploring the relevance of practice theories for the governance of sustainable consumption practices in the new world order.' *Global Environmental Change* 21(2011): 813–822. www.estudosdoconsumo.com.br/wp-content/uploads/2012/09/Gert-theories-of-practices.pdf (accessed 28 September 2015).

Teng, P. and S. Trethewie. 2012. *Tackling Urban and Rural Food Wastage in Southeast Asia: Issues and Interventions*. NTS Policy Brief No. 17. Singapore: RSIS Centre for Non-Traditional Security (NTS) Studies. http://www3.ntu.edu.sg/rsis/nts/HTML-Newsletter/PolicyBrief/pdf/Policy_Brief_280912.pdf (accessed 11 June 2015).

Wailes, E. and Chavez, E. 2012. *ASEAN and Global Rice Situation and Outlook*. ADB Sustainable Development Working Paper Series. www.adb.org/sites/default/files/publication/29969/adb-wp-22-asean-global-rice-situation.pdf (accessed 11 June 2015).

Whitehair, K., C. Shanklin and L. Brannon. 2013. 'Written messages improve edible food waste behaviors in a university dining facility.' *Journal of the Academy of Nutrition and Dietetics* 113(1): 63–69. www.sciencedirect.com/science/article/pii/S2212267212016425 (accessed 20 September 2015).

Williamson, D., H. Allen, P. Martin, A. Alfonso, B. Gerald and A. Hunt. 2003. 'Comparison of digital photography to weighed and visual estimation of portion sizes.' *Journal of the American Dietetic Association* 103(9): 1139–1145. www.sciencedirect.com/science/article/pii/S0002822305015464 (accessed 20 September 2015).

Index

affluence 1, 40, 125, 169, 173, 200
Age Good Food Guide 127–128
agriculture 46, 47, 49, 50–56, 62, 72, 107–108, 110–113, 119, 120, 171, 197, 199, 216; organic agriculture or organic farming *see* organic
air-conditioning 2, 77, 99, 102–104
alternative food movements, *see* food
altruism 9, 196
animal: feed 7, 14, 48, 49, 56, 57, 59, 61, 168; products 1, 27, 155, 175; *see also* meat *and* fish
Asia, Asia and the Pacific, Asian cities or countries 2, 5, 6, 11, 12, 13, 14, 16, 23, 27, 41, 46, 47, 50, 52, 53, 55, 58, 59, 61, 62, 72, 74, 97, 109, 118, 131, 125, 196, 227
Asia Pacific 125
Asian: consumers 109; cuisine 201; migrants 131; *see also* Asian cities
Asians 115
aquatic products 28, 31, 32, 33, 34, 35; *see also* fish
audit *see* waste audit
Australia 2, 54, 124, 125, 126, 131, 132, 136, 137
authentic 152
authenticity 108, 113, 114, 127, 129
avoidable food waste *see* waste

Bangalore 15, 92, 96; *see also* Bengaluru
Beef *see* meat *and* culture (meat)
Bengaluru 6, 53, 57, 58, 59–60, 141–155
bio-economics 3; *see also* Georgescu-Roegen
bio-gasification 6, 14
bio-methanation 59
biophysical: activity 1, 63; considerations or perspectives 2, 174; dimension 3–4, 6, 10; flows 203; patterns 9, 16, 211; studies 232; understanding 5, 14, 62, 240
bins *see* waste bins
Bourdieu, P. 4, 11, 71, 72, 73, 78, 182

café culture *see* culture
canteen 10, 12–14, 59, 60, 77, 87, 146, 174, 190, 239–250; *see also* food service sector
caste or caste society 5, 8, 9, 11, 29, 141–147, 149, 151–156
casteism 150–151
chicken *see* poultry
China 2, 6, 9, 23, 24–28, 30–31, 34–35, 37–39, 42, 90, 109, 110, 112, 127, 131, 141, 180–197, 239
Chinatown 133 (Melbourne)
city 2, 5, 14, 31, 33, 34, 35, 53, 57, 58, 62, 69, 85, 101, 103, 108, 124–125, 146, 172, 191, 197; Bengaluru 151; Ho Chi Minh City 71; Honolulu 115; Kyoto City 175; Matsumoto City 171; Melbourne 127–138; Shanghai as a global city 181; *see also* Bangalore or Bengaluru, Honolulu, Ho Chi Minh, Kyoto, Melbourne, Metro Manila, Shanghai, *and* mega city
class 5, 11, 73, 78, 90, 93, 95, 155, 194; creative class 129; working class 93, 99; *see* also status and middle classes
climate change 1, 50, 61, 119, 216
coffee 73–75, 78–80, 82–83, 96, 102, 107, 111, 112, 127–128, 133–134, 136, 150
colonial or colonialism 74, 75, 93, 98, 110, 136, 142, 238
composting *see* waste compost
Community Supported Agriculture (CSA) 52, 166
condominiums 7, 12, 13, 90–105

consumer citizens 2, 9, 196
consumerism 7, 28, 62, 71–72, 86–87, 196
convenience 2, 5, 10, 12, 80, 98, 100, 171, 218–219, 225, 229, 230, 231, 232, 233; convenience store 96, 222–224, 227; convenience vegetables 228, 230; convenient food 12, 36, 227, 228; quest for convenience 12
cultural capital 128, 137, 180
culture 5, 6, 7, 8, 23, 24, 31, 32, 38, 61, 62, 82, 90, 124, 127, 129, 131, 133, 134, 137, 149, 151, 194, 205, 218, 224, 225; café, bar or restaurant culture 8, 12, 79, 124, 129, 133, 134, 138; counter culture or counter cultures 7, 85, 87; consumer or consumption culture 28, 35, 91, 195, 217; Dalit 152; food culture 1, 13, 31, 73, 74, 81, 112, 132, 183, 194, 238; global consumer culture 13, 72, 77, 78, 83, 86; Hawaiian and Hawaiian indigenous 113; Indian 153; material culture 4; meat culture 28, 32; Spam culture 98; youth culture 74, 83, 145

dairy or dairy products 38, 111, 126, 142, 163, 164, 202, 208, 210; *see also* milk
developing countries 1, 3, 6, 23–25, 27, 34, 40, 47, 50, 58, 62, 199, 239
diaspora or diasporas 7, 8, 125, 127, 131–134
đổi mới 72, 74, 75, 78–80, 83

eating out 2, 6, 7, 11, 12, 13, 36, 37, 52, 71–87, 100–101, 135, 162, 164–165, 167, 195, 200, 228; *see also* food service sector
eco-efficiency 207, 209, 210
ecological economics 3
ecological footprint 6, 28, 41
economic flows *see* flows
ecosystem services 1, 3
education or educational 6, 11, 28, 32, 55, 74, 75, 81, 109, 114, 116, 121, 135, 142, 167, 185, 188, 193, 227, 232, 239, 240, 241, 248, 249
eggs 7, 11, 27, 29, 38, 42, 114, 115, 116, 117, 118 , 126, 133, 136, 142, 148–151, 154, 156, 162, 163, 184, 192, 202, 210, 222, 246; eggetarian 149; egg shells 201
electricity 13, 47, 48–49, 97, 99, 100, 102, 103, 104, 119
elite 11, 38, 41, 53, 75, 91
empirical 2, 11, 15, 23, 137, 182, 195, 201, 203, 244

end of pipe approach 3
environmental impacts 1, 2, 3, 4, 6, 14, 46, 47, 50, 51, 52, 54, 56, 60, 61, 62, 171, 174, 182, 231, 243, 247
European Commission 55
expenditure 34, 37, 41, 42, 75, 164, 165, 197

fair trade 108
famine 5, 9, 175, 180, 183, 188, 189, 195
fast food *see* food
fertilizer 48, 50, 51, 52, 53, 55, 154, 192, 193; organic 46, 192
fish 6, 27, 29, 31, 32, 33, 34, 38, 39, 42, 110, 111, 113, 142, 147, 148, 152, 169, 171, 176, 186, 193, 194, 197, 202, 209, 210, 241, 242, 244, 245, 246
fishing 116, 167
flows: economic flows 202, 203, 210; food flows 10, 93, 231, 248; material flow analysis (MFA) 4; material, resource or biophysical flows 3, 4, 47, 173, 175, 62, 203, 208, 209, 250; migration or migratory flows 5, 11; tourism flows 137
food: alternative food networks (AFNs) or movements 107, 108–110; banks 9, 14, 74, 84, 85, 87, 172, 173, 174, 176, 233; certification 54, 55, 119; chains (local and global) 1, 77, 100, 147, 148, 166–167; court 96, 100, 222, 239; fast food, 2, 7, 8, 13, 38, 40, 71, 73, 74, 75, 76, 77, 78, 80, 82, 83, 96, 135, 138, 144, 164; foodscape 74, 108, 110; healthy 108, 136; identities 7, 108, 117–118; imported or imports 54, 114, 137, 154; junk food 38; labelling 121, 144; leftover 59, 60, 102, 104; local 7, 8, 11, 13, 107–121, 161, 166, 167, 171–174, 184, 189, 227, 238; losses 1, 208, 239; miles or mileage 8, 108, 166; production 1, 3, 6, 16, 46, 50, 51, 62, 108, 114, 119, 136, 138; processed or packaged 1, 60, 61, 75, 98, 113, 121, 144; provisioning 1, 183, 203, 208, 211, 231; retail or retailers, stores, coops 10, 76, 107, 108, 117, 132; safety 38, 119, 166, 211, 212, 213, 219, 230, 242, 244, 250; salvage parties 9, 14, 172–174; scandals 38; scarcity 11, 74, 85, 86, 180, 195; security 61, 216; service sector 9, 199–250; sharing 1, 7, 9, 11, 71–87, 146, 147, 150, 166, 169, 174, 184–186, 195; sovereignty 108, 113, 114, 117; surplus 74, 85, 98, 162, 169, 172, 199, 200, 206, 211, 212, 213, 233;

256 Index

system 1, 5, 7, 13, 62, 71–74, 80, 86, 114, 168, 173, 199, 233, 250; unhealthy 114, 147; unsafe 109; *see also* convenient food *see* convenience; food culture *see* culture; food flows *see* flows; food practices *see* social practices; vegetarian food *see* vegetarian; wasted food *see* waste

Food and Agriculture Organisation (FAO) 24, 36, 42, 141, 155

foodie 8, 76, 124, 125, 128, 129, 130, 137

friendship 9, 141, 142, 145, 146, 147, 154, 155

frugality 9, 13, 79, 181, 186, 188, 195–196

garden 125, 193; home garden 137; kitchen garden 52; school garden program 121

gender or gendered practices 7, 11, 13, 71, 73, 74, 77, 78–80, 81, 82, 86, 155, 162, 227

generational or inter-generational 7, 11, 12, 71, 73, 74, 78, 80–83, 86, 195

Georgescu-Roegen, N. 3

Giddens, A. 4, 72, 92, 182

global food chains *see* food

globalization 108, 182

globalized 7, 108, 120

Global North 6, 24, 28, 34, 40, 41, 142, 152

Global South 23–42, 232

green revolution 50, 53, 62, 141

grocery shopping 10, 216, 219, 220, 224, 225, 226, 229, 230

habitus 71, 73, 78; *see also* Bourdieu

Hawai'i 7, 8, 107–121

HCMC *see* Ho Chi Minh City

home production 137

Honolulu 2, 11, 107–121

Hosier Lane 128, 129, 130

hospitality 77, 80, 82, 83–84, 191

households 1, 2, 6, 8 10, 11, 12, 13, 16, 33, 34, 41, 42

household consumption expenditure *see* expenditure

Ho Chi Minh City 2, 7, 11, 71–87

hypermarket *see* food retail

identity 7, 11, 26, 71, 76, 83, 85–87, 91, 112, 153; middle-class 73; self-identity 91, 125, 137, 217

India 2, 6, 8, 11, 23–42, 47, 52, 53, 59, 62, 92, 96, 127, 141–156, 239

individual action framing or individualisation 1, 4, 174, 181, 200, 217, 221, 240, 241

industrial ecology 2, 3, 4, 6, 10, 202, 214, 239; industrial ecology group, University of Lausanne 5

industrial metabolism *see* metabolism

interviews 10, 74, 115, 125, 169, 182, 191, 216, 220, 231, 244–245; group interviews 161–162; household interviews 220; in-depth interviews 15, 201–202; informal interviews 9, 180–181, 242, 245, 247; key informant interviews 7, 92–93; oral history interviews 74; semi-structured interviews 201

interdisciplinary or interdisciplinarity 2, 3, 10, 14–15, 199, 201, 214

inter-generational *see* generational

International Federation of Organic Agriculture Movements (IFOAM) 50

interventions 10, 136, 213, 241, 245–249

Japan 2, 9, 14, 27, 110, 112, 161–175, 191

kitchen garden *see* garden

Kuala Lumpur 2, 6, 9, 12–14, 47, 50, 53–54, 59–60, 199–214, 239

landfill 7, 14, 46, 49, 56–61, 167, 201, 216, 220

leftovers *see* food leftovers

liberalization 1, 221

licensing 60, 129–130, 133

Life Cycle Assessment (LCA) or analysis 4, 6, 46, 56, 62, 243

Life Cycle Inventors (LCI) 47, 55

life cycle thinking or considerations 4, 6, 46–47, 56, 61, 14

lifestyles 6, 9, 11, 28, 40, 73, 113, 114, 170, 199, 225, 232, 233; consumer 229; middle-class 76; modern 165; sustainable 196; Western 221;

local food *see* food

Malaysia 2, 9, 15, 47, 52, 54, 58, 60, 110, 199–214, 232

mall 2, 12, 76, 90, 91, 95, 96, 99, 100, 102, 104, 250

Manila *see* Metro Manila

market liberalization *see* liberalization

material flows or material flow analysis (MFA) *see* flows

meat 2, 4, 9, 24–35, 38–40, 50, 54, 59, 98, 99, 111, 116, 119, 126, 136, 141, 142,

144, 146–148, 150–151, 153–156, 162–164, 183, 186, 188, 190, 192–193, 202, 208, 210, 229, 241–246; meat consumption 6, 8, 11, 14, 23–41, 50, 141–145, 153–155, 192
mega city 39, 146; Hyderabad 147; Metro Manila 7, 90–106; Shanghai 9
Melbourne 2, 8, 11–13, 71, 124–140
metabolism 3, 4, 6
Metro Manila 2, 6, 7, 10–14, 47, 50, 53, 54, 58–60, 90–106, 238–250
MFA *see* flows
middle-class/middle classes 6; middle-classness 7, 11, 73, 75, 82, 84, 86, 87; *see also* new middle classes
migrants or migration 11
Millennium Consumption Goals 41
Ministry of Agriculture, Forestry and Fisheries (Japan) 165

national data 2
National Project on Organic Farming (India) 52, 53
new consumers 6, 7, 72
new middle classes 6
New Organic Regulations (South Korea) 55
non-vegetarian *see* vegetarian

Organisation for Economic Cooperation and Development (OECD) 24–26, 34
organic: agriculture or farming, 6, 14, 46, 47, 50–55, 172; farms and farming 51–54, 118, 172; fertilizer *see* fertilizer; food and produce 11, 46, 50, 52–54, 55, 109–110, 137; Organic Agriculture Act (Philippines) 54; Organic World Congress 55; production 11, 53, 108; stores and coops 52, 107, 108; waste *see* waste

peri-urban 136; fringe 138; kitchens 52
Philippines 2, 7, 15, 47, 52, 54, 55, 60, 90, 91, 92, 94, 101, 102, 112, 116, 232, 238, 250
poor 23, 27, 33, 39, 41, 60, 85, 113, 114, 116, 180, 191, 194
poultry 24, 26, 27, 29, 30, 34, 35, 39, 50, 126, 136, 241, 242, 244–246
power 5, 111, 144, 187; purchasing power 28, 32, 40–42, 63, 75, 86; rate 97; structure 229; supply 59
practices 1–16, 28, 30, 41, 46, 50, 51, 52, 53, 55, 58, 59, 60, 62, 63, 71, 72, 73, 74, 76, 77, 78, 80, 81, 82, 83, 84, 85, 86, 87, 90, 91, 92, 93, 98, 100, 101, 104, 107, 108, 110, 119, 125, 141, 142, 143, 144, 145, 146, 147, 148, 150, 151, 153, 154, 155, 161, 162, 165, 168, 169, 170, 171, 174, 175, 176, 180, 181, 182, 184, 187, 188, 194, 195, 197, 199, 200, 203, 205, 206, 210, 211, 212, 213, 214, 216–220, 223, 225, 228, 229, 230, 231, 232, 233, 239, 240–241, 244, 246, 247, 248, 249, 250
practice theory *see* social practice theory

Reckwitz, A. 4, 91, 182, 217, 240, 241
reflexivity 15
refrigerators or refrigeration 5, 59, 99, 101, 102, 162, 218, 219, 227, 229, 231
restaurant 2, 4, 8, 9, 10, 11, 12, 13, 16, 38, 52, 53, 60, 61, 74, 75, 76, 77, 80, 82, 83, 85, 96, 99, 100, 102, 103, 107, 113, 124, 127, 128, 130, 132, 133, 134, 135, 137, 138, 144, 147, 150, 152, 164, 165, 166, 167, 168, 171, 173, 174, 181, 184, 185, 190, 193, 194, 195, 199, 200, 201, 202, 203, 204, 206, 208, 210, 211, 212, 213, 220, 239; *see also* food service sector
retail *see* food retail
rice 10, 13, 14, 54, 55, 83, 98, 102, 111–113, 133, 136, 161–164, 170, 175, 176, 180, 184, 205, 207, 238–250
rich 39–41, 180, 194
rural 2, 24, 28–42, 53, 75, 94, 100, 108, 113, 162, 180, 189, 190, 197

salvage parties *see* food salvage parties
Schatzki, T. 4, 28, 91, 92
school 81, 108, 121, 136, 161, 162, 165, 170, 174, 176, 186, 188–191, 193, 238–250
Seoul 2, 6, 10, 12, 14, 16, 47, 50, 53–61
service sector *see* food service sector
Shanghai 2, 9, 13, 81, 109, 175, 180–197
sharing *see* food sharing
Shove, E. 4, 91, 92, 182, 217, 218, 219, 221, 232, 240
social: distinction *see also* Bourdieu and class; inequality 7, 11, 16, 72, 85–87; justice 108, 120; learning 14, 171, 241, 247, 249, 250; norms 2, 5, 10, 11, 15, 29, 30, 73, 77–80, 82, 84, 86, 92, 173, 182, 206, 212, 213, 218, 232, 239, 242, 246, 250; practices see *practices*
social practice theory, theories or approach 4–6, 9, 10, 11, 16, 72, 91–92, 180, 182, 203, 214, 240–241, 249

social status *see* status
South Asia 152
Southeast Asia 46, 62, 74, 90
South Korea 2, 47, 55, 61
Spam 98, 113; *see also* culture (Spam culture)
status 4, 5, 11, 16, 38, 59, 73, 75, 77, 78, 82, 83, 84, 86, 137, 142, 144, 187, 217; socio-economic status 134; *see also* class
students 10, 241, 242–250; adolescent students 135; university or college students 74, 77, 79, 80, 82, 86, 141, 142, 146, 185, 233
suburban 124, 136, 171
supermarketization 221; *see also* food retail
sustainability 7, 8, 13, 23, 27, 46, 63, 85–87, 108–109, 113–114, 120, 141, 154, 155, 166, 180–197, 217, 239, 249
sustainable food consumption 3, 6, 9, 14, 23, 62, 180, 195, 200, 239, 248, 249
systems of provision 5, 9, 10, 12, 16, 36, 165, 166, 175, 216–233

table manners 80, 180
thrift 9, 13, 180–197
time: scarcity 10, 217, 221, 228; use, 216, 220–226
trans-disciplinary or trans-disciplinarity 14–16
transport or transportation 1, 50, 56, 57, 61, 95, 103, 104, 136, 138, 181, 184, 218, 219, 230, 243
Tokyo 2, 161–176
tourism 2, 94, 112, 125, 134, 137
tourists 8, 38, 113, 124, 128, 181, 203, 205
toxicity 61
trash bins *see* waste bins

unavoidable food waste *see* waste
university 5, 10, 74, 77, 79, 99, 112, 120, 146, 182, 185, 186, 229, 233, 241, 243–247, 250; *see also* schools
urban or urban areas 2, 5, 7–9, 11–13, 24, 28, 30–42, 46, 52, 53, 56, 59, 60, 71–76, 85–87, 90, 91, 94, 95, 100, 109, 116, 118, 125, 128, 130, 138, 147, 166, 171, 180, 189, 190, 195–197, 223, 230, 232, 239
urbanization 1, 155, 180, 182, 216, 221, 226, 232
USA 2, 25, 26, 38, 54, 55, 60, 108, 109, 134, 162

vegetarian or vegetarianism 5, 8, 9, 11, 28–30, 38, 40, 42, 83, 141–156, 192–193; non-veg 8, 9, 11, 141–156
vertical neighbourhoods 7, 13, 90–104
Vietnam 2, 7, 11, 12, 71–87, 127

Warde, A. 2, 4, 5, 40, 72, 91, 92, 176, 182, 217, 218
Waste: audit 202, 241, 242, 246, 249, 250; bins 244; compost or composting 7; food waste 1, 2, 4, 6, 9, 10, 12, 13, 14, 15, 46–49, 56–62, 72, 74, 81, 85, 86, 87, 92, 161–250; incineration 7, 14, 46, 49, 56–61; organic 56, 57, 101, 102; prevention 10, 199, 200, 202, 206, 212–214; segregation 57–60, 101, 220, 242; unavoidable 244
World Organization of Organic Farms (WOOF) Network 118

youth 8, 9, 11, 74, 82, 83, 114, 141–156, 162